AN OBSERVER'S GUIDE TO THE EARTH'S MOON

A Lunar Atlas For Amateur Astronomers

Maynard Pittendreigh

Mustard Seed Books
Brunswick GA
1996

ISBN: 978-0-6151-3528-1

For more information, visit www.Pittendreigh.com.

3

TABLE OF CONTENTS

Introduction 5

Observing the Moon with the Naked Eye 7

Lunar Features to Observe with Binoculars 25

Craters of the 4-Day Moon – Binoculars 35

Craters of the 7-Day Moon – Binoculars 53

Craters of the 10-Day Moon – Binoculars 83

Craters of the 14-Day Moon – Binoculars 115

Lunar Features – Visible with a Telescope 121

Craters of the 4-Day Moon – Telescope 147

Craters of the 7-Day Moon – Telescope 161

Craters of the 10-Day Moon – Telescope 173

Craters of the 14-Day Moon – Telescope 189

Lunar Eclipses......................................195

A Lunar Glossary 207

Introduction

This book offers the serious amateur astronomer a simple lunar atlas built around the Astronomy League's Lunar Observing Program.

The League's observing programs offer encouragement and structure to amateurs who wish to be challenged to develop their knowledge and skill in observational astronomy. The Lunar Program introduces amateur astronomers to an object in the sky that many of us take for granted, and for which deep sky observers despise. The program allows amateurs in heavily light polluted areas to participate in an observing program of their own. The program is perfect for the beginning and inexperienced observer, as well as the experienced amateur who has spent most of his or her time looking at more distant objects.

The program requires a person to observe 100 features on the moon -- 18 naked eye, 46 binocular, and 36 telescopic features. Any pair of binoculars and any telescope may be used for this program. Observing the objects in the program, and in this atlas, does not require any expensive equipment.

OBSERVING THE MOON WITH THE NAKED EYE

8

Old Moon in the New Moon's Arms

And

New Moon in the Old Moon's Arms

The dark area of the lunar disk is sometimes illuminated by the reflected light from earth. If this comes within 72 hours prior to the new moon, it is called "the old moon in the new moon's arms." If it comes 72 hours after the new moon, the effect is referred to as "the new moon in the old moon's arms."

Imaginary Images On The Moon

Astronomers detest a full moon. The bright moon blocks out many deep sky objects elsewhere in the sky, and the glare and lack of shadows on the surface of the moon make lunar observation difficult. It might be interesting, however, for you to go out on an evening of a full moon to see if you can locate some of the traditional images on the moon.

The **"man on the moon"** (left) may be the most visible illusion seen on the face of a full moon. Find the two eyes, and the rest can more easily be "seen" with just a little bit of imagination.

The **woman on the moon** (right) takes just a wee bit more imagination. Find her flowing hair, then the outline of her face, and the mind should then "see" the woman's image.

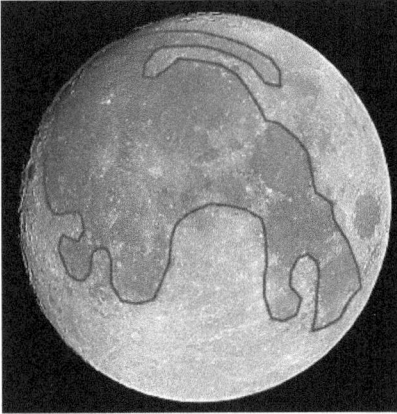

Is there a rabbit **jumping over the moon?** The photo on the left is the outline of what many see when they view the moon – a long eared rabbit. Don't try to find the bunny tail, however. It's not there.

The last photograph shows **the cow jumping over the moon.** It is almost the same as the rabbit. The cow can best be viewed shortly before the full moon. Some see a cow's headless body with four legs. Others see a three legged cow with the head grazing on the moon (the head would be the extension on the far right of the drawing).

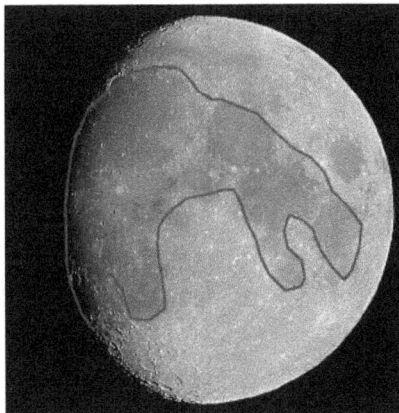

MARIA

The Lunar **maria** (singular: **mare)** are large, dark, basaltic plains on Earth's Moon, formed by ancient basaltic flood eruptions caused by extremely large meteoroid impacts. They were named dubbed maria (which is Latin for "seas") by early astronomers who mistook them for actual seas.

Maria are less reflective than the 'highlands' or mountains, which are older and have had a soil or 'regolith' created by the impact of micro-meteoroids over hundreds of millions of years. The regolith is more reflective than the basalt of the maria.

The traditional nomenclature for the Moon also includes one "oceanus" (ocean), as well as features with the names "lacus" (lake), "palus" (marsh) and "sinus" (bay). The latter three are smaller than maria, but have the same nature and characteristics.

There are a number of Maria which can often be located with the naked-eye: **Crisium, Fecunditatis, Frigoris, Imbrium, Nectaris, Nubium, Humorum, and Oceanus Procellarum**, **Serenitatis and Tranquillitatis.**

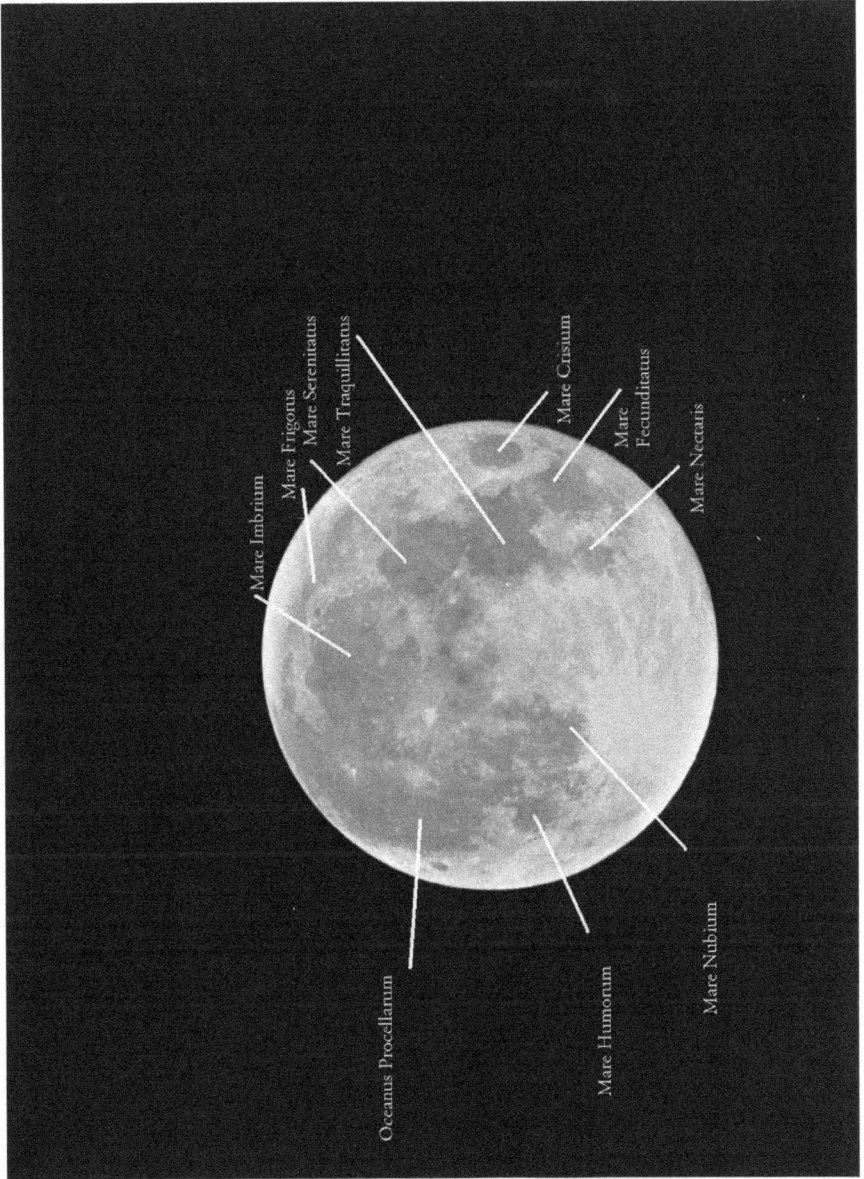

Crisium

The Mare Crisium is more commonly known as "the sea of crises." It is 350 miles in diameter in the east-to-west direction and 270 miles in the north-to-south direction. It covers an area of 66,000 square miles. Crisium is near the eastern edge of the side of the moon visible from Earth and forms one of the more striking features of the lunar surface. Mare Crisium is the setting for Arthur C. Clarke's 1951 short story, "The Sentinel."

Mare Crisium

Fecunditatis

Mare Fecunditatis is more commonly called the "Sea of Fecundity" or "Sea of Fertility.") It is a lunar mare 909 miles in diameter. This basin is overlapped with the Nectaris, Tranquillitatis, and Crisium basins. Fecunditatis basin meets Nectaris basin along Fecunditatis' western edge. On the eastern edge of Fecunditatis is the crater Langrenus.

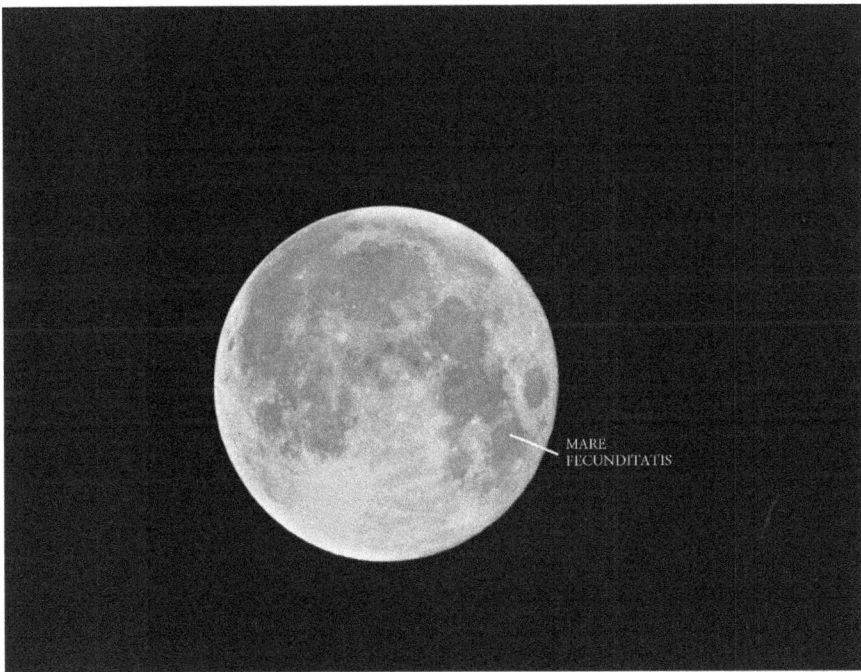

MARE
FECUNDITATIS

Serenitatis

Mare Serenitatis is more commonly known as the "sea of serenity." It is a lunar mare that sits just to the east of Mare Imbrium. Both Luna 21 and Apollo 17 landed near the east border of Mare Serenitatis, in the area of the Montes Taurus range.

Some people believe Mare Serenitatis is one of the eyes for the Man in the Moon.

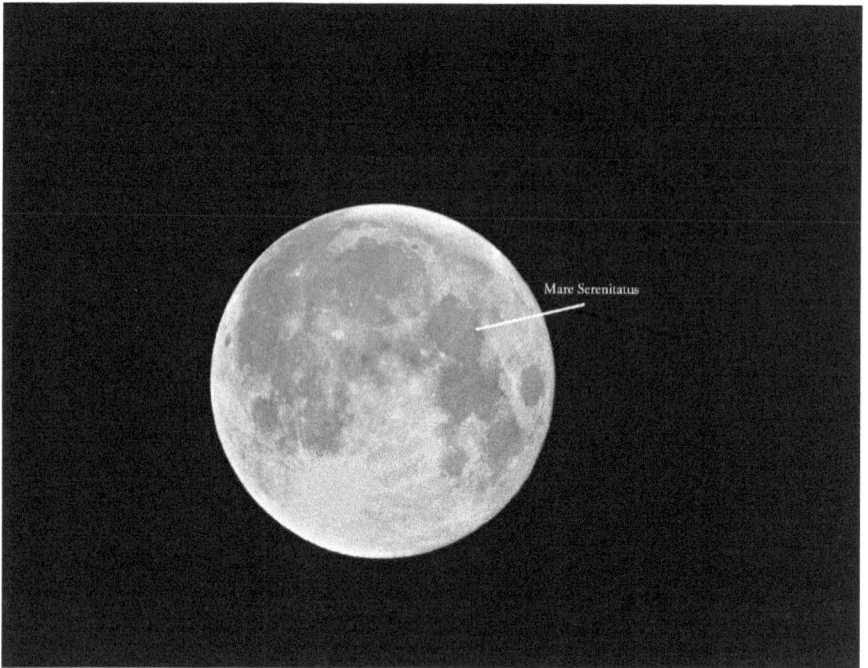

Mare Serenitatus

Tranquillitatis

Mare Tranquillitatis is more commonly called "sea of tranquility." In 1965, the Ranger 8 spacecraft crashed in Mare Tranquillitatis, after successfully transmitting 7,137 photographs of the moon in the final 23 minutes of its mission. This mare also served as the landing site for the Apollo 11 lunar module, the first manned landing on the Moon. The landing area at 0.8° N, 23.5° E has been designated Station Tranquillitatis, and three small craters to the north of the base have been named Aldrin, Collins and Armstrong in honor of the Apollo 11 astronauts.

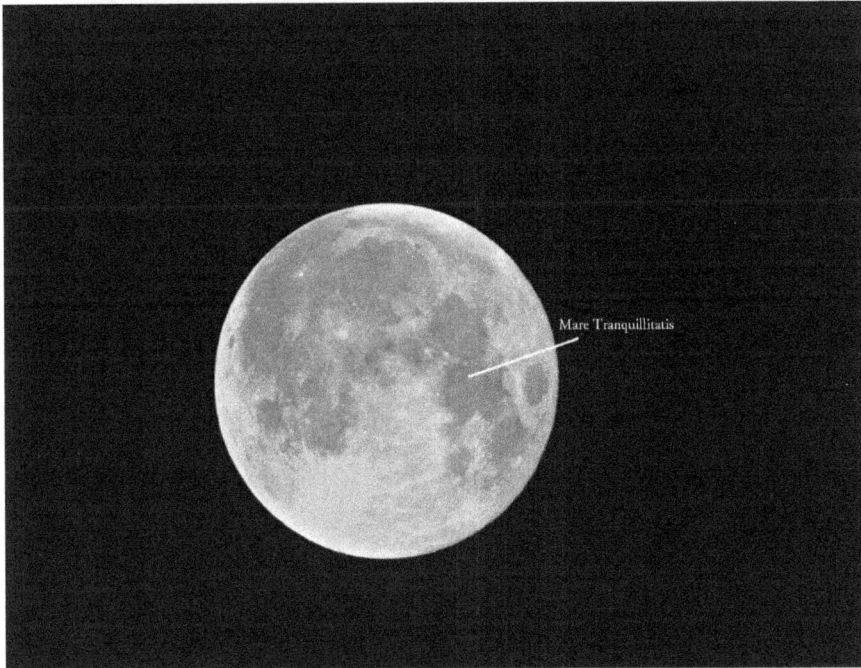

Mare Tranquillitatis

Nectaris

Mare Nectaris is more commonly known as the "sea of nectar." It is a rather small lunar mare or sea (a volcanic lava plain noticeably darker than the rest of the moon's surface) located somewhat between the Sea of Tranquility (Mare Tranquillitatis) and the Sea of Fecundity (Mare Fecunditatis). Montes Pyrenaeus borders the mare to the west and the large crater near the south center of the mare is known as Rosse.

Mare Nectaris

Imbrium

Mare Imbrium is more commonly known as the "Sea of Showers" or "Sea of Rains." With a diameter of 1123 km it is second only to Oceanus Procellarum in size among the maria, and it is the largest mare associated with an impact basin. Apollo 15 landed in the southwestern region of Mare Imbrium, near the Apennine Mountains.

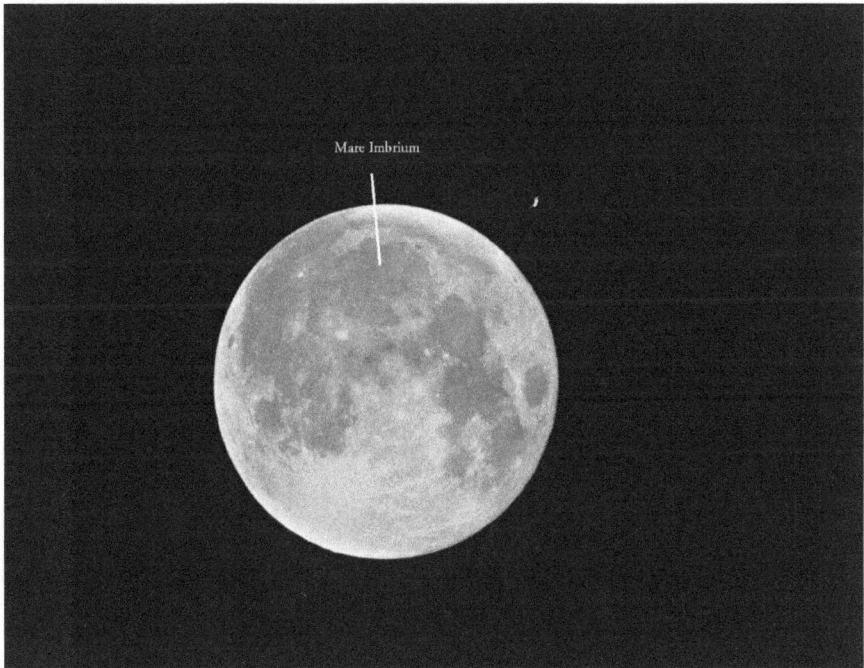

Mare Imbrium

Frigoris

Mare Frigoris is more commonly known as the "Frigid Sea" or the "Sea of Cold." It is a lunar mare located just north of Mare Imbrium, and stretches east to north of Mare Serenitatis. The dark circular feature just to the south of Frigoris is Plato crater.

Nubium

Mare Nubium is more commonly known as the "Sea of Clouds." It is a lunar mare located just to the southeast of Oceanus Procellarum.

Mare Nubium

Humorum

Mare Humorum is more commonly known as the "Sea of Moisture." It was not sampled by the Apollo program, so a precise age has not been determined. However, geologic mapping indicates that it is intermediate in age between the Imbrium and Nectaris Basins, suggesting an age of about 3.9 billion years. On the north edge of Mare Humorum is the large crater Gassendi, which was considered as a possible landing site for Apollo 17.

Mare Humorum

Oceanus Procellarum

Oceanus Procellarum is more commonly known as "Ocean of Storms." It is a vast lunar mare on the western edge of the near side of Earth's Moon. Its name derives from the old superstition that its appearance during the second quarter heralded bad weather. Procellarum is the largest of the lunar maria.

The unmanned lunar probes Surveyor 1, Surveyor 3, Luna 9 and Luna 13 landed in Oceanus Procellarum. Apollo 12 also landed in Oceanus Procellarum.

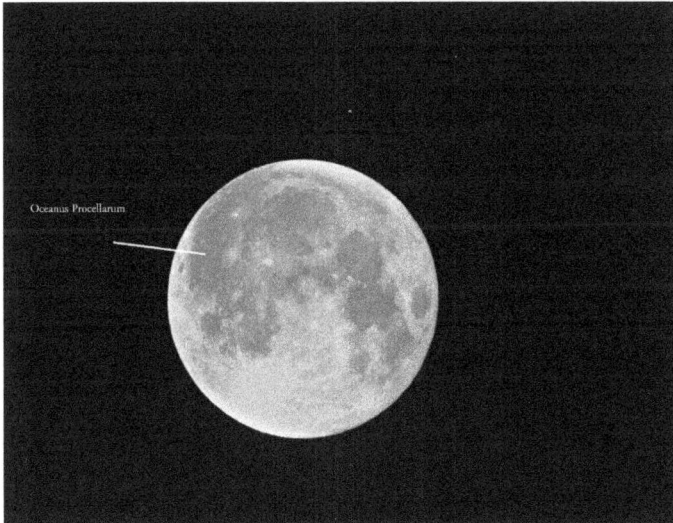

LUNAR FEATURES TO OBSERVE WITH BINOCULARS

Lunar Rays

Lunar rays are streaks of fine ejecta thrown out during the formation of an impact crater. The photos below show the crater, Tycho. Although the rays are very clear and prominent, the second photo highlights the locations of some of the rays of Tycho.

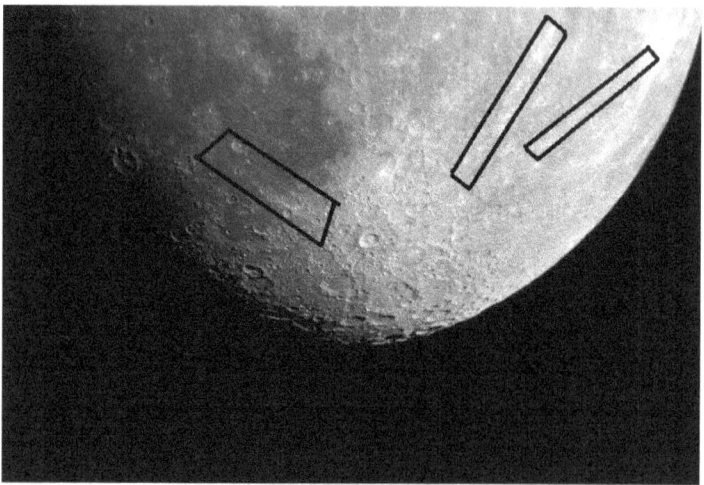

Ray systems were once thought to be only found on planetary bodies that lack an atmosphere, but have since been discovered on Mars. Ray ejecta material have different reflectivities (i.e., aldebo), compositions. Typically visible rays have a higher albedo than the surrounding surface, making them stand out as brighter features that can form streaks and patterns across the surface. More rarely an impact will excavate low albedo material.

Among the lunar craters on the near side with pronounced ray systems are Aristarchus, Copernicus, Kepler, Proclus, and Tycho.

Sinus Iridum

Sinus Iridum is more commonly called the "Rainbow Sea." It is a plain of basaltic-lava that forms a northwestern extension to the Mare Imbrium. The feature was given the Latin name for the Bay of Rainbows by Giovanni Riccioli.

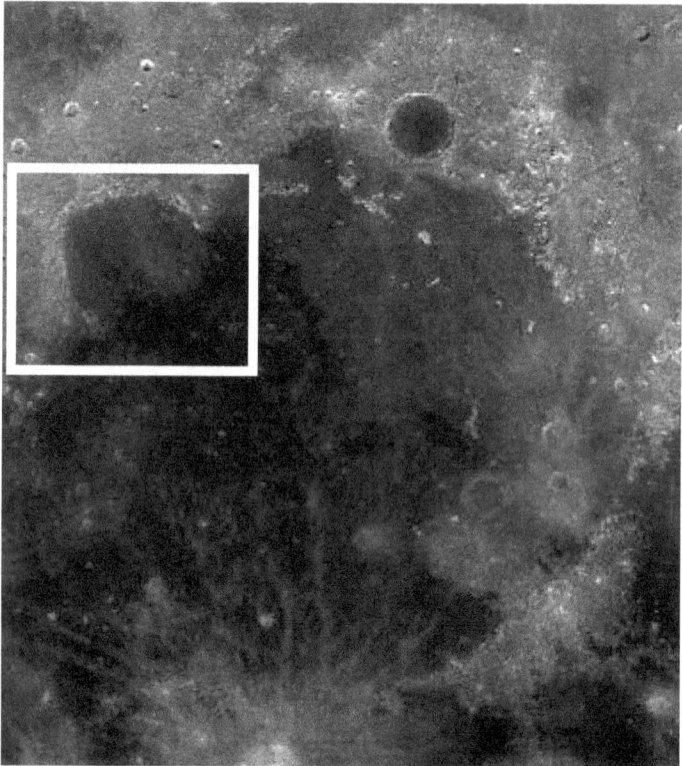

To Locate: Find Mar Imbrium. Sinus Iridum is the "bay" that connects with Imbrium.

Sinus Medii

Sinus Medii is a small lunar mare that is located at the intersection of the Moon's equator and prime meridian. As seen from the Earth, Sinus Medii is located in the central part of the Moon's near side, and it is the point closest to the Earth. This plane was given the Latin name for "Central Bay' by Johannes Mädler.

Sinus Medii

To Locate: The observer should find the crater Copernicus, highlighted in the photograph by the square box. Then located the dark areas that form a letter "C." Sinus Medii is the southern leg of the letter "C." And alternative approach, if Copernicus is not visible, is to look for Mare Serenitatis and Mare Tranquillitatis, and then look south for the "C."

Sinus Roris

A northward extension of the Oceanus Procellarum has been given the Latin name for "Bay of Dew". It is a rather long area, and the borders of this feature are somewhat indistinct.

To Locate: Find Mare Imbrium and Oceanus Procellarum. Look between them, and toward the north for a smaller dark area.

Palus Somnii

The common name for this area is "the Sea of Sleep."

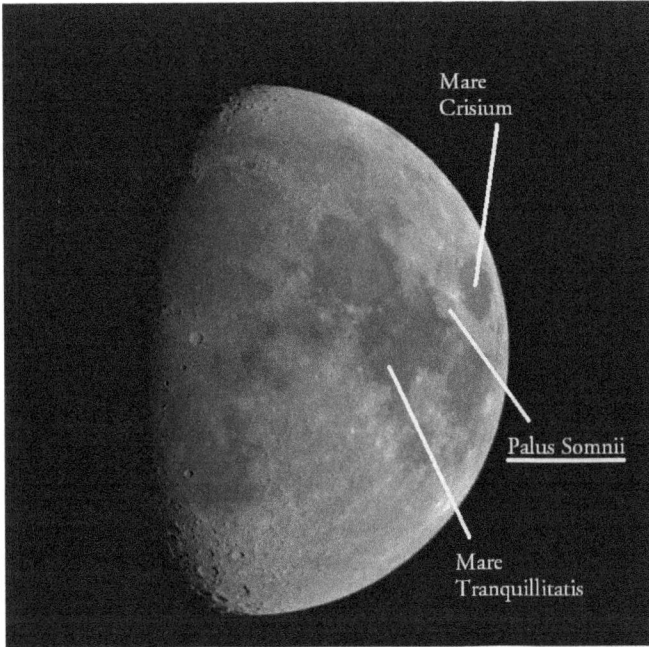

Mare Crisium

Palus Somnii

Mare Tranquillitatis

To Locate: Find Mare Crisium and Mare Tranquillitatis. Palus Somnii is an area between the two that is of a unique brightness – not as dark as a Mare, not as bright as the rest of the moon.

Palus Epidemiarum

Palus Epidemiarum is more commonly called, the "Marsh of Epidemics." It is a small lunar mare in the southwestern part of the Moon's near side. It lies to the southwest of Mare Nubium, and southeast of Mare Humorum. This feature forms a rough band of lava-flooded terrain that runs generally west-east, with a northward extension near the western end.

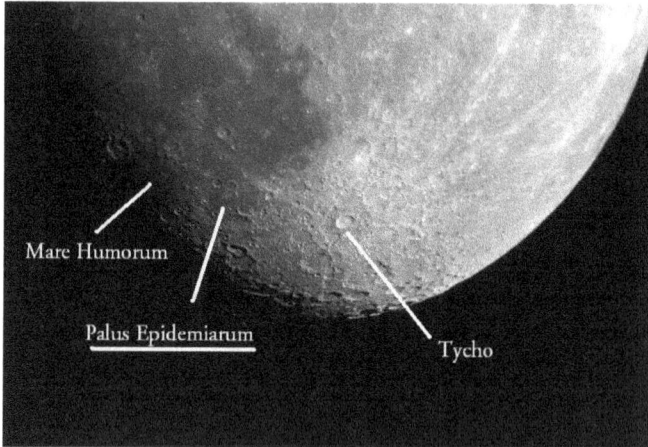

Mare Humorum

Palus Epidemiarum

Tycho

To Locate: Find the bright rayed crater, Tycho. Along one direction of the rays are areas of maria. Palus Epidemiarum is a small mare-like area between Tycho and Mare Humorum.

Mare Vaporum

Mare Vaporum (the "sea of vapors") is a lunar mare located between the southwest rim of Mare Serenitatis and the southeast rim of Mare Imbrium. The lunar material surrounding the mare is from the Lower Imbrian epoch, and the mare material is from the Eratosthenian epoch. The mare lies in an old basin or crater that is within the Procellarum basin. To the south of the mare is a light colored thin line. This feature is Rima Hyginus. The mare is bordered by the mountain range Montes Apenninus.

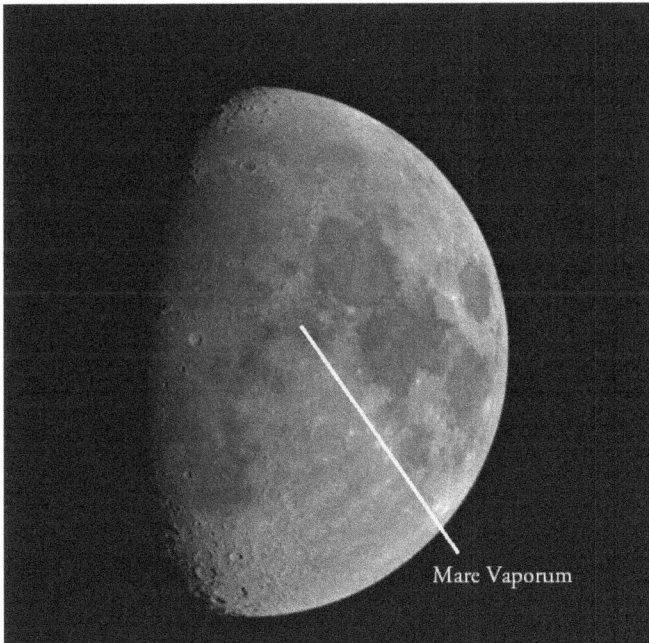

Mare Vaporum

To Locate: Vaporum is a small mare – look in the area surrounded by Mare Tranquillitatis, Mare Imbrium and Mare Serenitatis.

CRATERS OF THE FOUR DAY OLD MOON

Visible with Binoculars

Langrenus

Langrenus is a prominent impact crater located near the eastern lunar limb. The feature is circular in shape, but appears oblong from earth. It lies on the eastern shore of the Mare Fecunditatis.

The interior of the crater has a higher albedo than the surroundings, so the crater stands out prominently when the sun is overhead.

During the Apollo 8 mission, Astronaut James Lovell described this feature as "quite a huge crater; it's got a central cone to it. The walls of the crater are terraced, about six or seven terraces on the way down."

The Flemish Astronomer Michel Florent van Langren was the first person to draw a lunar map while giving names to many of the features. He even named this crater after himself. Ironically, this is the only one of his named features that has retained his original designation.

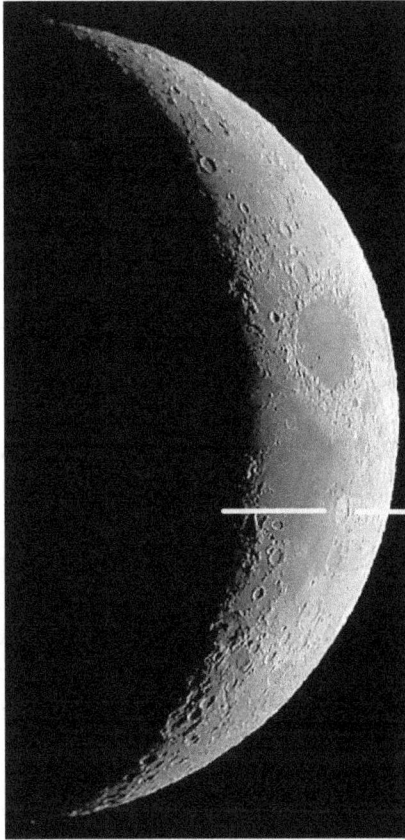

TO LOCATE: On a 4 day moon, there will be two maria visible – Crisium and Foecunditatis. Crisium is a distinctly separate maria. Foecunditatis will be near the terminator. Crater Langrenus will be on the edge of Foecunditatis, toward the lunar limb.

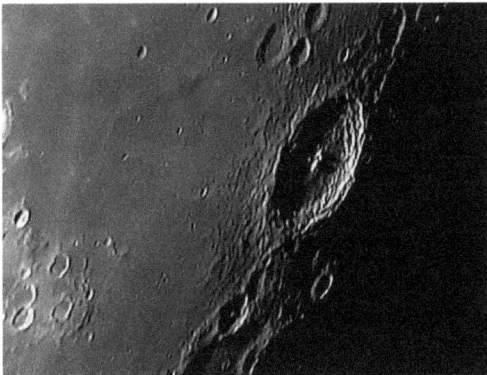

Vendelinus

Vendelinus is a lunar impact crater located on the eastern edge of Mare Fecunditatis. To the north of Vendelinus is the prominent Langrenus crater, while to the southeast is Petavius crater; forming a chain of prominent craters near the eastern rim. From the vantage point of Earth, the crater appears oblong due.

The crater is heavily worn and overlapped by multiple craters, making this feature more difficult to identify except at low sun angles. The crater's irregular rim is broken in several places by overlapping craters. The most prominent of these is the break in the northeast wall from the overlapping Lamé crater. The smaller Lohse crater overlaps the rim to the northwest, and at the south end the crater wall is joined to the Holden crater.

The floor of Vendelinus is flat and covered by a dark lava flow. Webb noted that a dark patch could be seen on the floor of this crater at full moon. Vendelinus lacks a central peak, but includes multiple impact craters of various dimensions.

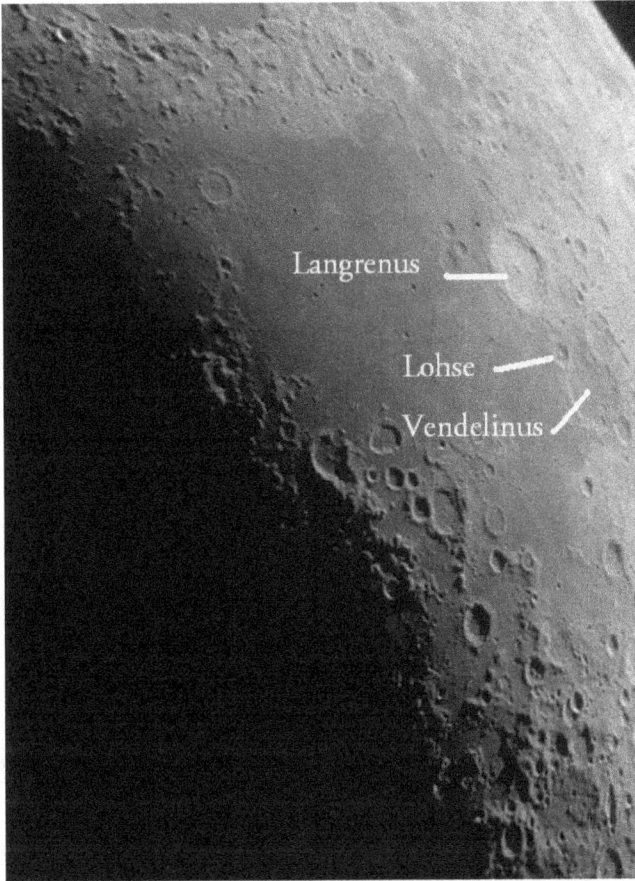

TO LOCATE: Langrenus is a bright crater with a central peak. From there, identify a crater of equal size, but not as bright. This crater will be intruded upon but a somewhat smaller crater (Lame) and surrounded by two much smaller craters that touch its rim. That shallow, dimmer, crater equal in size to Langrenus is Vendelinus.

Petavius

Petavius is a large lunar impact crater located to the southeast of the Mare Fecunditatis, near the southeastern lunar limb. The crater Petavius appears oblong when viewed from the Earth due to foreshortening.

The outer wall of Petavius crater is unusually wide in proportion to the diameter. It displays a double-rim along the south and west sides. The crater floor displays a rille system named the Rimae Petavius. A deep fracture runs from the peaks toward the southwest rim of the crater.

The Reverend T. W. Webb described Petavius as "one of the finest spots in the Moon: its grand double rampart, on east side nearly 11,000 ft. High, its terraces, and convex interior with central hill and cleft, compose a magnificent landscape in the lunar morning or evening, entirely vanishing beneath a Sun risen but halfway to the meridian."

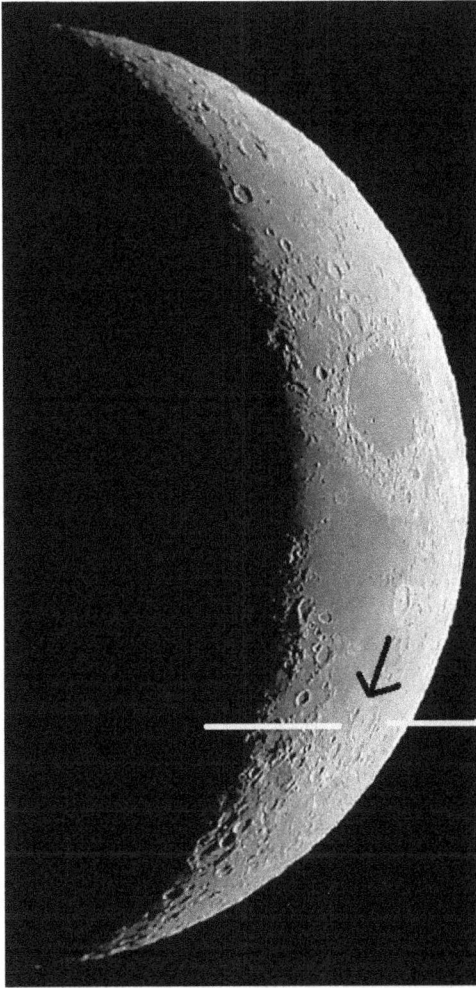

TO LOCATE: From Langrenus, move southward toward Petavius. The most favorable time for viewing this feature through a telescope is when the Moon is only three days old. By the fourth day the crater is nearly devoid of shadow.

Cleomedes

Cleomedes is a prominent crater located north of Mare Crisium. It is surrounded by multiple crater impacts and rough ground. The irregular Tralles crater intrudes into the northwest rim. Delmotte crater is to the east. A triple-crater formation is located north of Cleomeses, with Burckhardt crater in its center.

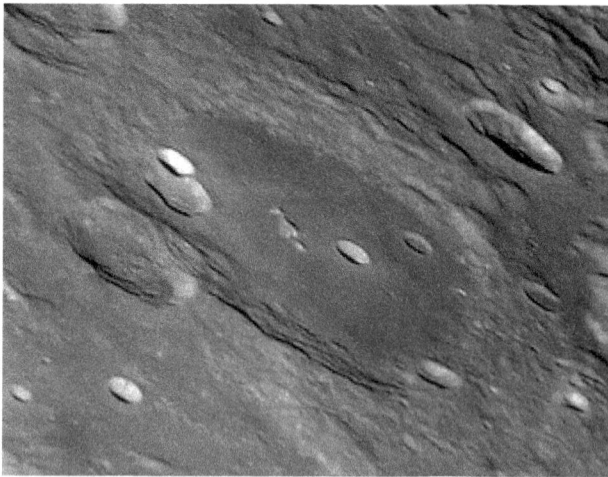

The outer wall of this crater is heavily worn, especially along the southern part of the wall. The crater floor is nearly flat, with a small central peak to the north of the mid-point. There are several notable craterlets on the floor, including a pair of overlapping craters just inside the northwest rim.

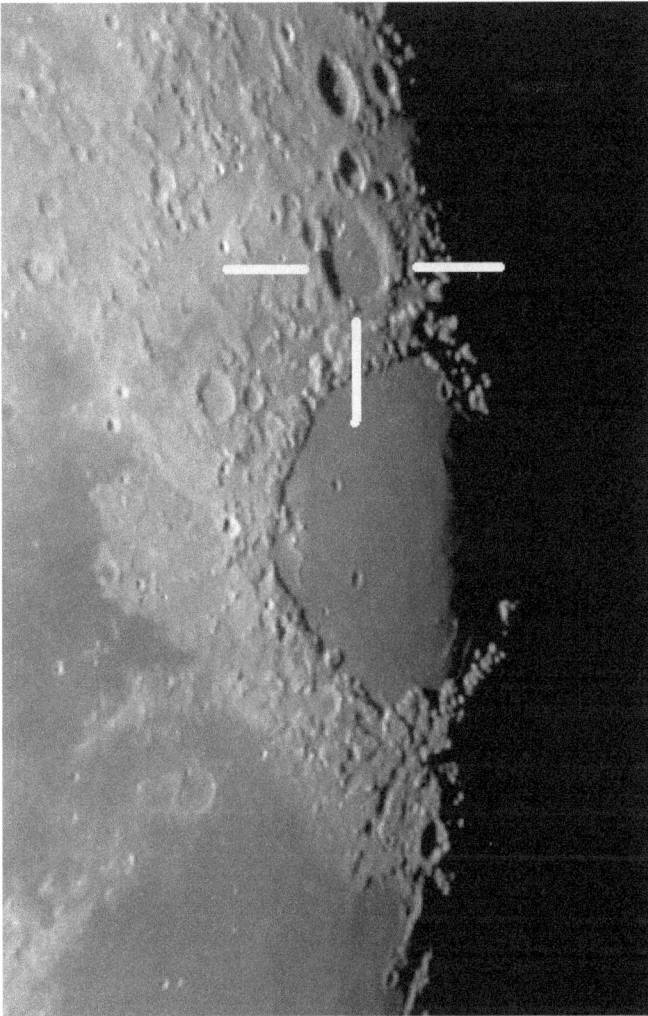

TO LOCATE: Find Mare Crisium and look slightly toward the lunar North.

Atlas

Atlas is a prominent impact crater located in the northeast part of the Moon, to the southeast of Mare Frigoris. The somewhat smaller but more prominent Hercules crater is to the west.

The inner wall of the Atlas crater multiply terraced and the edge slumped, forming a sharp-edged lip. Atlas is a floor-fractured crater with a rough and hilly interior that has a lighter albedo than the surroundings. Floor-fractures are usually created as a result of volcanic modifications.

There are two dark patches along the inner edge of the walls; one along the north edge and another besides the southeast edges. A system of slender clefts named the Rimae Atlas crosses the crater floor, and were created by volcanism. Along the north and northeastern inner sides are a handful of dark-halo craters, most likely formed due to eruptions. Around the mid-point is a cluster of low central hills arranged in a circular formation.

TO LOCATE: Atlas will be along the lunar terminator when the moon is at its 4th day.

Hercules

Hercules is a prominent crater located in the northwest part of the Moon, to the east of Atlas crater. It lies along the east edge of a southward extension in the Mare Frigoris. It is next to the larger Atlas Crater.

The interior walls of Hercules have multiple terraces. There is a small outer rampart.

The crater floor has been flooded by lava in the distant past, and contains several areas of low albedo. The central peak has been buried, leaving only a low hill near the mid-point.

The satellite crater 'Hercules G' crater is located prominently just to the south of the center. The small crater 'Hercules E' lies along the southern rim of Hercules.

Locate

TO LOCATE: At the 4th day, Hercules will be next to Atlas, but a bit more into the shadow of the terminator.

Endymion

Endymion is a crater that lies near the northeast limb of the Moon. The Atlas crater is located to the southwest. From the view point of Earth, the Endymion crater has an oval appearance.

The floor of Endymion has been covered in low albedo lava, giving it a dark appearance and makes it relatively easy to locate. The floor is almost smooth and featureless, with a few tiny craterlets located within the rim. A string of three such craters are found near the northwestern inner wall. Faint streaks of ray material from Thales crater to the north-northwest crosses the dark floor.

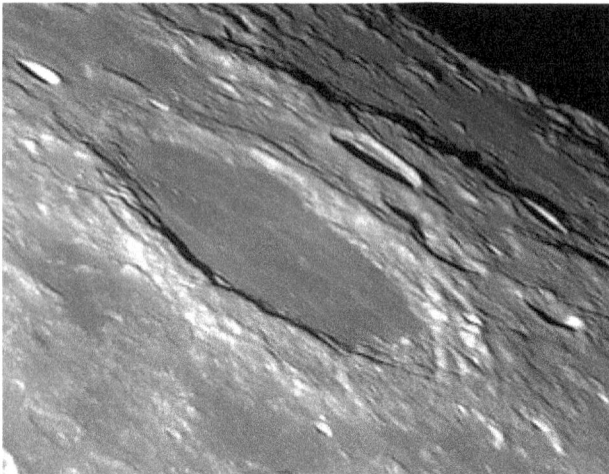

TO LOCATE: If you have already located Atlas and Hercules, then simply look toward the limb for a dark crater.

Macrobius

Macrobius is an impact crater located to the northwest of the Mare Crisium. Tisserand, a somewhat smaller crater, lies just to the east. Tisserand and Macrobius make a notable pairing. The lines in the photograph below point to Macrobius. With the smaller Tisserand, the pair looks somewhat like a figure "8."

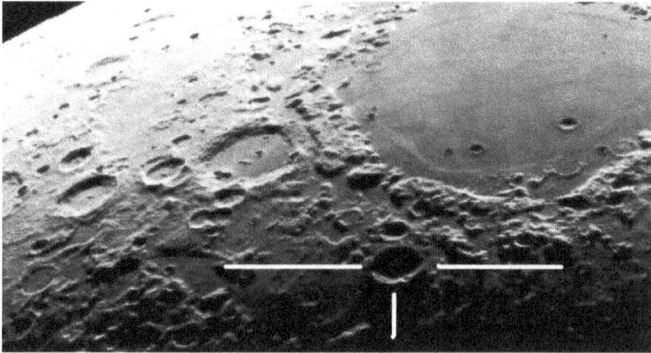

The outer wall of Macrobius has a multiply-terraced inner surface. There is some slumping along the top of the rim.

The small satellite crater 'Macrobius C' is located across the western rim, but the wall is otherwise relatively free of significant wear. There is a central mountain complex in the center of the crater floor. There is a low ridge in the western interior, but the remainder of the floor is relatively level.

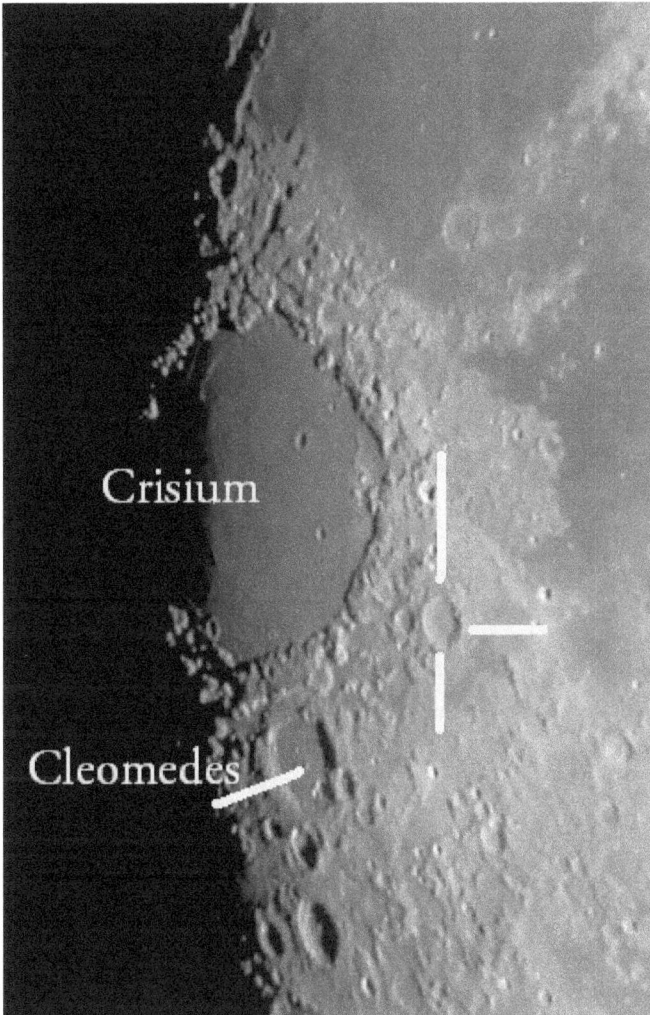

Crisium

Cleomedes

TO LOCATE: Find Crisium. Tisserand and Macrobius are a figure "8" pair of craters just beyond the edge of Crisium.

CRATERS OF THE SEVEN DAY OLD MOON

Visible with Binoculars

Piccolomini

Piccolomini is a prominent impact crater located in the southeastern sector of the Moon.

Piccolomini's rim has not been severely worn by crater impacts. The inner wall possesses wide terraces. These structures have been somewhat smoothed by landslips and erosion, probably caused by seismic activity. An influx of material has entered across the northern rim, flowing down toward the base. The crater floor is relatively smooth, with only minor hills and impact craters. In the middle is a complex central peak surrounded by lesser mounts.

Rupes Altai is a lengthy escarpment in the lunar surface that begins at the western rim of Piccolomini crater, curving to the northwest.

TO LOCATE: Locate Mare Nectaris. Piccolomini will be a prominent crater with a central peak south of Nectaris.

Theophilus

Theophilus is a prominent impact crater between Sinus Asperitatis and Mare Nectaris. It partially intrudes into Cyrillus, a similar sized crater to the southwest. The Rev. T. W. Webb described Theophilus as "the deepest of all visible craters". The rim of Theophilus crater has a wide, terraced inner surface that indicates landslips.

The Apollo 16 mission collected several pieces of basalt that are believed to be ejecta from the formation of the Theophilus crater.

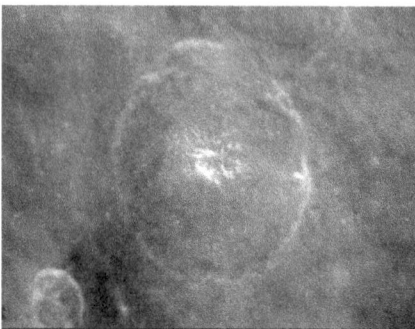

As seen in this image, when the moon is full and the sun is shining directly over the crater, the features seen in the top photograph are washed out.

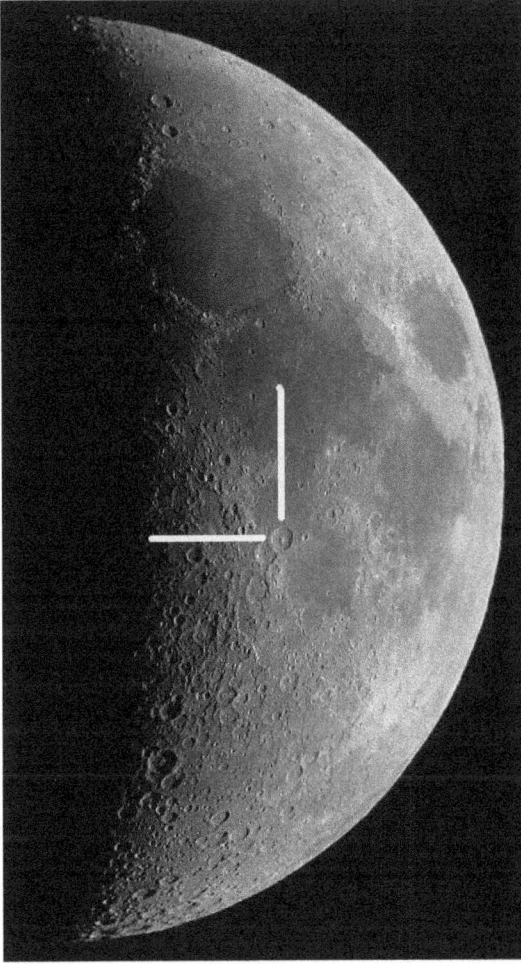

TO LOCATE: Locate Mare Nectaris.
Theophilus will be a prominent crater with a central
that is on the very outside edge of Nectaris.

Cyrillus

Cyrillus is an impact crater located on the northwest edge of Mare Nectaris. The equally large, but younger Theophilus crater intrudes upon Cyrillus.

The rim of Cyrillus crater has been worn, although it remains intact except where overlaid by Theophilus crater. The floor is rough and irregular, with three distinct central peaks offset slightly to the northeast.

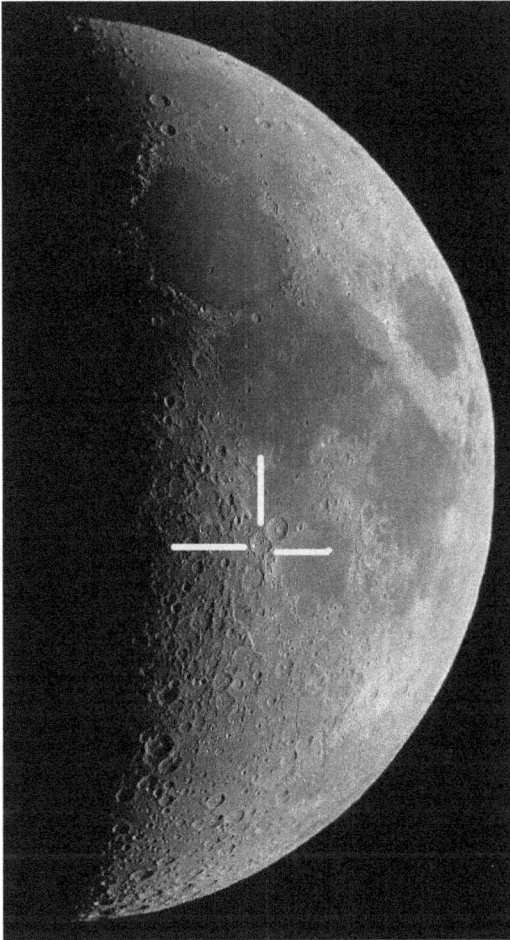

TO LOCATE: If you located Theophilus, Cyrillus is very easy. The crater that Theophilus intrudes upon is Cyrillus.

Catharina

Catharina is an ancient lunar impact crater located in the southern highlands. With the large Cyrillus and Theophilus craters in the north, it forms a prominent grouping that is framed by the curve of the Rupes Altai. These three structures form a wonderful view when the sun is at a low angle to the surface. There is also a distinct difference in the ages of these three craters, with the age increasing significantly from north to south.

The rim of Catharina is heavily worn and irregular, with most of the north wall incised by the worn ring of crater Catharina P. The northeast wall is deeply impacted by several smaller craters. No terracing remains on the inner wall, and the outer rampart has been nearly eroded away. The floor is relatively flat but rugged, with a curved ridge formed by Catharina P, and the remains of a smaller crater near the south wall. Nothing remains of a central peak.

The crater, Catharina, is named after Saint Catherine of Alexandria. Theophilus and Cyrillus are also named for Christian saints who lived in Alexandria.

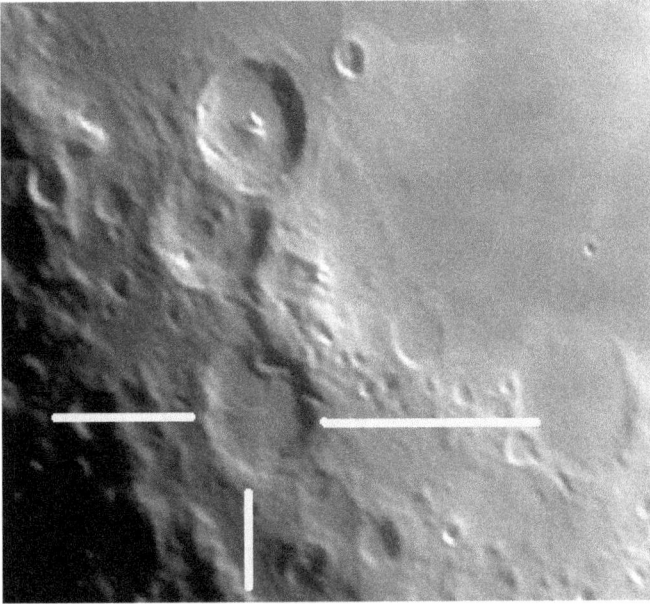

TO LOCATE: Once you find Theophilus and Cyrillus, Catharina will not be difficult. These three form a readily identifiable triad, with Theophilus in the north, just outside Mare Nectaris, then Cyrillus overlapping Theophilus. Catharina is the next in line, and is removed from the other three.

Posidonius

Posidonius is a lunar impact crater that is located on the western edge of Mare Serenitatis, to the south of Lacus Somniorum. The Chacornac crater is attached to the southeast rim The Daniell crater is to the north.

The rim of Posidonius is shallow and obscured, especially on the western edge. The interior has been overlaid by a lava flow in the past.

There is no central peak, but the floor is hilly and laced with a rille system named the Rimae Posidonius. There is a slight bulging in the crater floor due to the past lava uplift, which probably produced the complex of rilles. The northeast rim is interrupted by the smaller crater 'Posidonius B'. Within the crater rim is another smaller crater 'Posidonius A'.

The crater is named after Posidonius of Rhodes (or alternatively, "of Apameia"). He was a Greek Stoic philosopher, politician and astronomer of the 1st and 2nd Century BC. He was acclaimed as the greatest polymath of his age. None of his vast body of work can be read in its entirety today as it exists only in fragments.

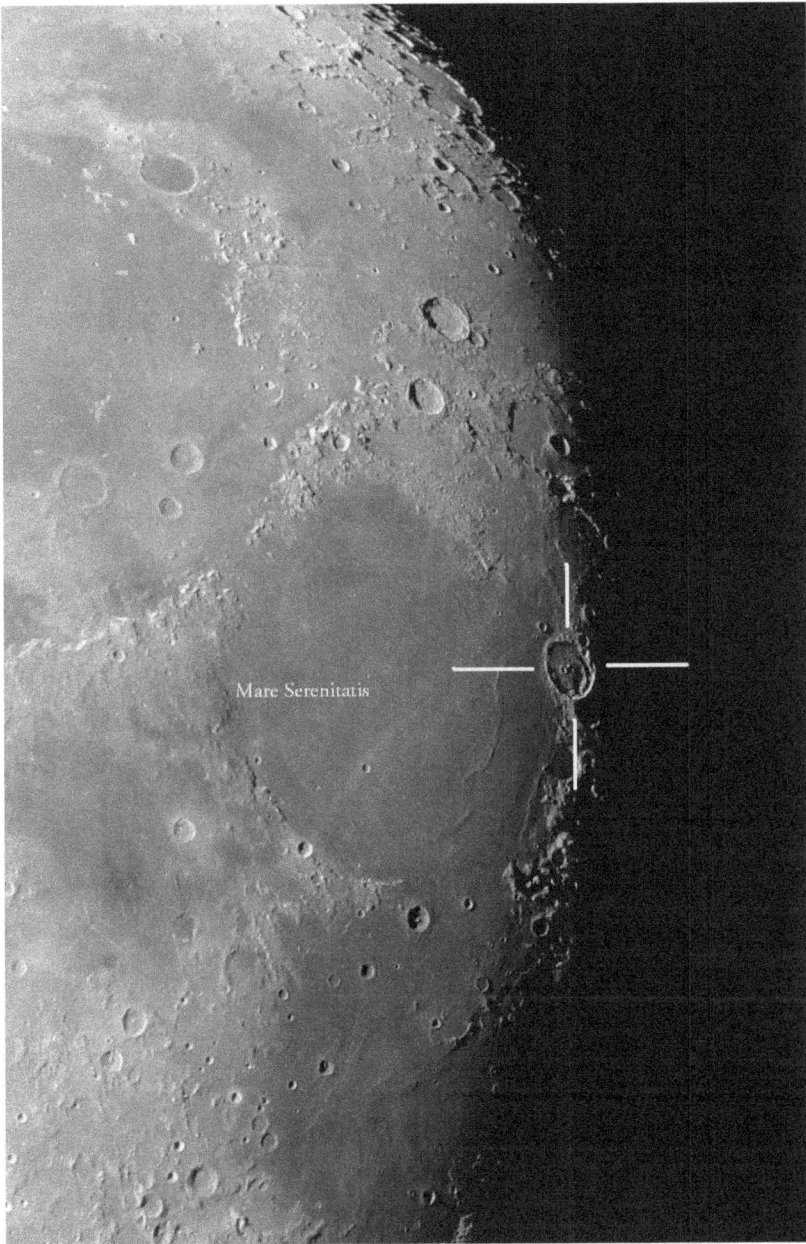

Mare Serenitatis

TO LOCATE: Posidonius is a prominent crater on the edge of Mare Serenitatis.

Fracastorius

Fracastorius is the lava-flooded remnant of an ancient impact crater located at the southern edge of Mare Nectaris.

The northern wall of this crater is missing, with only mounds appearing in the lunar mare to mark the outline. The lava that formed Mare Nectaris invaded this crater, so the structure now forms a bay-like extension. The remainder of the rim is heavily worn and covered in lesser impact craters, leaving little of the original rim intact. The most prominent of these craters is 'Fracastorius D', which overlays a portion of the western rim. The Fracastorius crater has no central peak.

The crater is named after Girolamo Fracastoro (Fracastorius) (1478-August 8, 1553), who was an Italian physician, poet, and scholar (in mathematics, geography and astronomy).

TO LOCATE: Locate Mare Nectaris. The "C" shaped crater intruding on the edge of the Mare is Fracastorius.

Aristoteles

Aristoteles is a lunar impact crater that lies near the southern edge of the Mare Frigoris. To the south of Aristoteles lies the slightly smaller crater Eudoxus. Aristoteles and Eudoxus form a distinctive pair for a telescope observer. The smaller Mitchell crater is directly attached to the eastern rim of Aristoteles. The inner walls are wide and finely terraced. The outer ramparts display a generally radial structure of hillocks through the extensive blanket of ejecta. The crater floor is uneven, and covered in hilly ripples. Aristoteles does possess central peaks, but they are somewhat offset to the south.

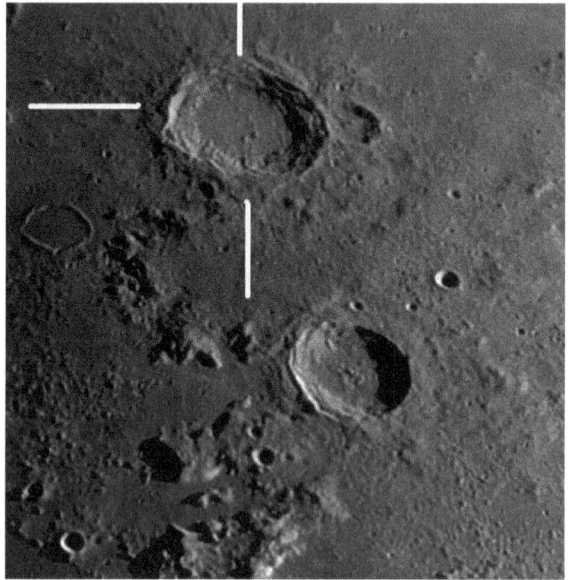

The crater is named for Aristotle, the great Greek philosopher who was a student of Plato and the teacher of Alexander the Great.

Mare Serenitatis

TO LOCATE: Locate Mare Serenitatis and move toward Mare Frigoris. Look past solitary crater, Eudoxus, and come to the slightly larger crater Aristoteles with its small partner hanging on the crater rim.

Eudoxus

Eudoxus is a prominent lunar impact crater that lies to the east of the northern tip of the Montes Caucasus range. It is located to the south of the prominent Aristoteles crater in the northern regions of the visible Moon. To the south is the ruined formation of Alexander crater, and the small Lamèch crater lies to the southwest.

The rim of Eudoxus has a series of terraces on the interior wall, and slightly worn ramparts about the exterior. It lacks a single central peak, but has a cluster of low hills about the mid-point of the floor. The remainder of the interior floor is relatively level.

The crater is named for Eudoxus of Cnidus (410 or 408 BC - 355 or 347 BC). Eudoxus was a Greek astronomer, mathematician, physician, scholar and friend of Plato. Since all his own works are lost, our knowledge of him is obtained from secondary sources, such as Aratus' poem on astronomy. In mathematical astronomy he is known for the introduction of the astronomical globe. He also made early contributions to understanding the movement of the planets.

Mare Serenitatis

TO LOCATE: Locate Mare Serenitatis and move toward Mare Frigoris. Don't move all the way to Aristoteles.

Cassini

Cassini is a lunar impact crater that is located in the Palus Nebularum, at the eastern end of Mare Imbrium. Promontorium Agassiz is to the northeast, the southern tip of the Montes Alpes mountain range. Theaetetus is to the south and Mons Piton is to the northwest.

The floor of the crater is flooded, and is likely as old as the surrounding maria. The surface is peppered with a multitude of impacts, including a pair of significant craters contained entirely within the rim. Cassini A is the larger of these two, and it lies just north-east of the crater center. A hilly ridge area runs from this inner crater toward the south-east. Near the south-west rim of Cassini is the smaller crater Cassini B.

The crater is named for Jacques Cassini and his father Giovanni Domenico Cassini. Giovanni Domenico Cassini (1625–1712) was an Italian astronomer, engineer, and astrologer. He discovered the Great Red Spot on Jupiter, four of Saturn's moons, Cassini Division in Saturn's ring system, and was the first to observe differential rotation within Jupiter's atmosphere.

Jacques Cassini (February 8, 1677 - April 18, 1756) was also a French astronomer who published the first tables of the satellites of Saturn in 1716.

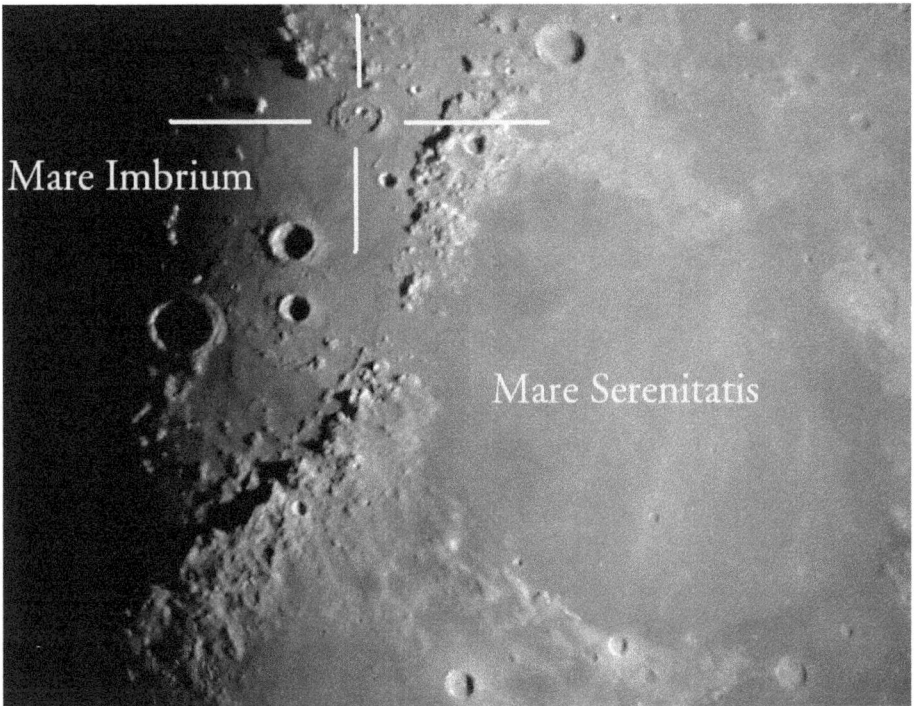

Mare Imbrium

Mare Serenitatis

TO LOCATE: Find Mare Serenitatis and Mare Imbrium. Look "just over the wall" from Serenitatis and there will be a crater with an off-center interior crater. That will be Cassini with the smaller Cassini A inside.

Hipparchus

Hipparchus is the degraded remnant of a lunar crater. It is located to the southeast of Sinus Medii, near the center of the Earth-view of the Moon. The Horrocks crater lies entirely within the northeast rim of the crater. The western rim has been all but worn away from impact erosion, and only low hills and rises in the surface remain to outline the feature. The wall to the east is intact, but it too is heavily worn. The crater floor has been partially resurfaced by basaltic lava flow. The southwest part of the floor, however, is slightly raised and much more rugged than the remainder. A few small rises and the raised partial rim of a flooded crater are all that remain of a central massif.

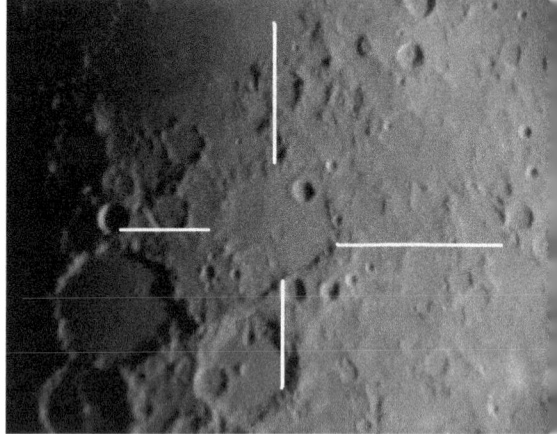

The crater is named for Hipparchus (ca. 190 BC–ca. 120 BC), a Greek astronomer. Hipparchus is considered the greatest astronomical observer, and by some the greatest astronomer of antiquity. He was the first Greek to develop accurate models for the motion of the Sun and Moon. His studies made it possible to predict solar eclipses. He compiled the first star catalogue of the western world.

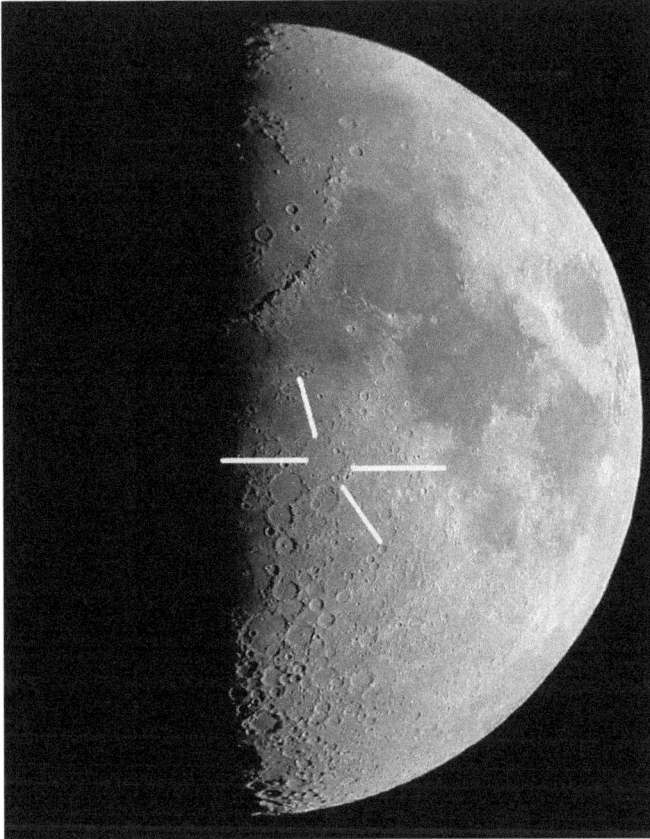

TO LOCATE: Hipparchus will be a bit of a challenge. It is best seen when it is along the terminator. At a quarter-moon when Crisium is visible, look along the terminator to a series of three large craters forming a slight curve – these will point mainly toward the north and slightly toward Crisium. These are (from south to north, smaller to larger) Arzachel, Alphonsus and Ptolemaeus. Hipparchus will be the slight remnant just north of Ptolemaeus.

Albategnius

Albategnius is an ancient impact crater located in the central highlands. The level interior is surrounded by a high, terraced rim. The outer wall is somewhat hexagon-shaped, and has been heavily eroded with impacts, valleys and landslips. The rim is broken in the southwest by the smaller Klein crater.

The crater is named for Al-Battani was an Arab Astronomer. His best-known achievement was the determination of the solar year as being 365 days, 5 hours, 46 minutes and 24 seconds.

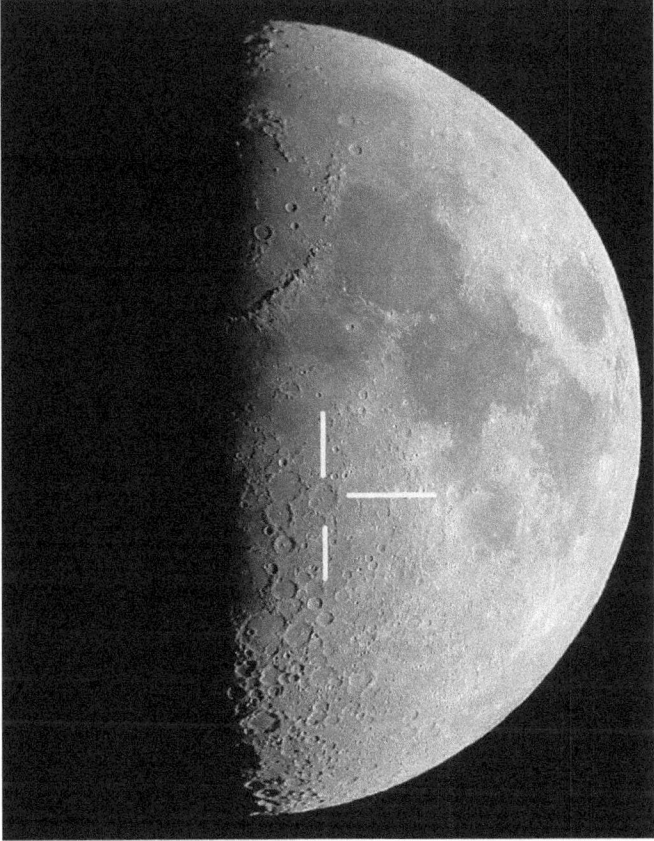

TO LOCATE: Albategnius can be found by finding the three craters Arzachel, Alphonsus and Ptolemaeus. Albategnius is prominent for its central peak and the small crater inside the main crater, along the rim.

Aristillus

Aristillus is a prominent impact crater that lies in the lunar mare at the southeast of Mare Imbrium. The Autolycus crater is directly to the south of Aristillus.

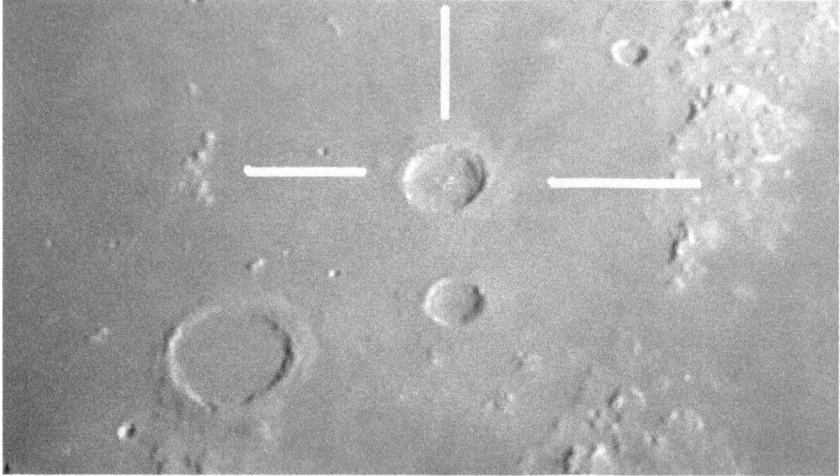

The rim of Aristillus has a wide, irregular outer rampart of ejecta that is relatively easy to discern against the smooth surface of the surrounding mare. The inner walls of the rim have a terraced surface, and descend to a relatively rough interior that has not been flooded with lava. In the middle of the crater is a set of three clustered peaks.

The crater was named after Aristillus, a Greek astronomer who created and early star catalogue in approximately 300 BC.

TO LOCATE: Locate the Mare Imbrium. Look toward the area of Imbrium that is close to Mare Serenitatis. In the area of three craters, Archimedes is the largest, and closer to the center of Imbrium; Autolycus is a smaller, smooth crater; and Aristillus is the middle sized of the three with a central peak and rough floor.

Autolycus

Autolycus is an impact crater that is located in the southeast part of Mare Imbrium. The Archimedes crater, which is more than double the size of Autolycus, is located to the west. Aristillus is located to the north.

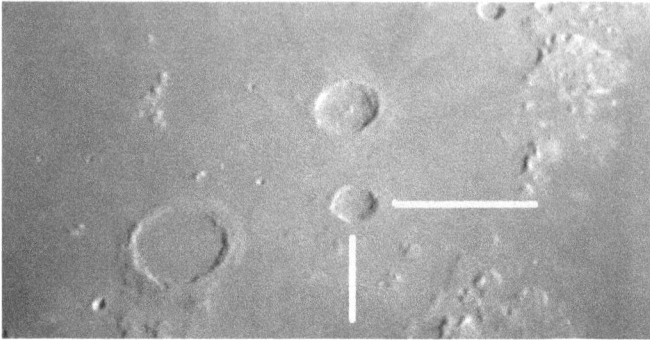

The rim of Autolycus is rather circular, but is somewhat irregular. It has a small outer rampart and an irregular interior with no central peak. It possesses a light ray system.

The Soviet's Luna 2 probe crash-landed just to the west-southwest of the crater rim.

The crater is named after Autolycus of Pitane (ca. 360 BC - ca. 290 BC), a Greek astronomer, mathematician, and geographer. In astronomy, Autolycus studied the relationship between the rising and the setting of the celestial bodies, and recorded his belief that "any star which rises and sets always rises and sets at the same point in the horizon."

TO LOCATE: Locate the Mare Imbrium. Look toward the area of Imbrium that is close to Mare Serenitatis. In the area of three craters, Archimedes, Autolycus and Aristillus, the smallest is Autolycus.

Maurolycus

Maurolycus is one of the more prominent lunar craters in the southern highland region of the Moon that is covered in overlapping crater impacts. It is challenging to find because it is in the midst of so many other craters. Just outside its rim is the slightly smaller figure "8" craters of Barocius and Barocius A. crater.

The outer wall of Maurolycus is tall, wide, and terraced.

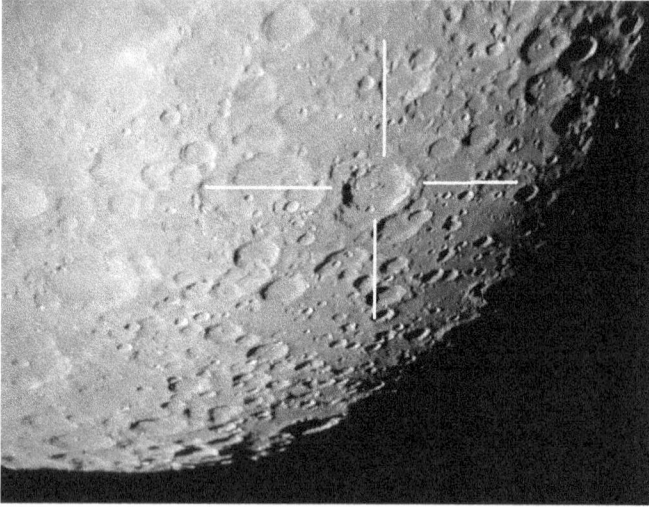

The crater is named for Francesco Maurolico (September 16, 1494-July 21 or July 22, 1575) was an Italian mathematician and astronomer.

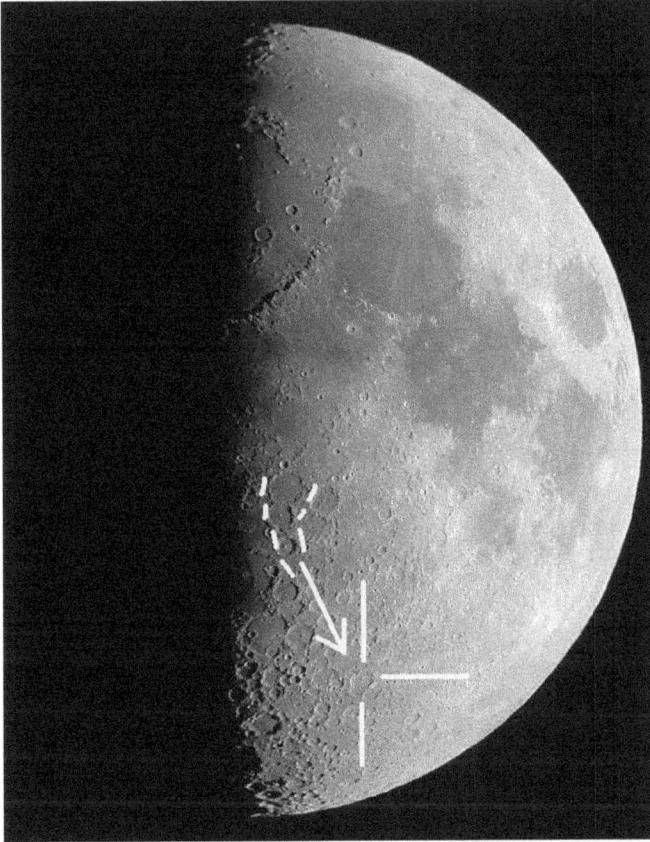

TO LOCATE: Maurolycus can sometimes be found along one of the rays of Tycho, if the sun's light is hitting it at a proper angle. The slight curve of three craters Ptolemaeus, Alphonsus and Arzachel give somewhat of a direction finder for Maurolycus.

CRATERS OF THE TEN DAY OLD MOON

Visible with Binoculars

Plato

Plato is the maria-surfaced remains of a lunar impact crater. It is located on the northeastern edge of the Mare Imbrium. Mare Frigoris is to the north.

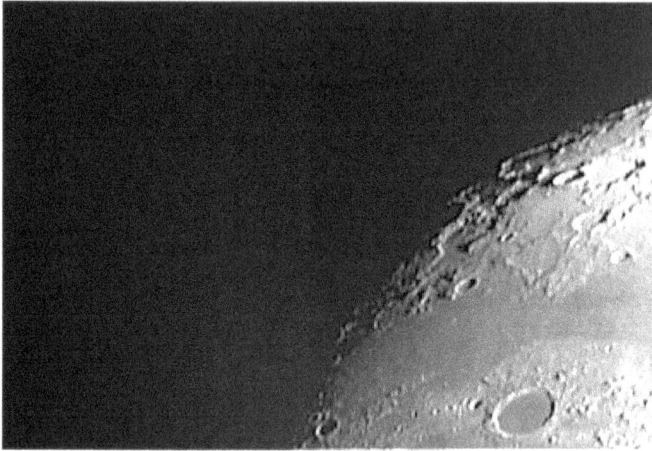

The rim is irregular with jagged peaks that project prominent shadows across the crater floor when the sun is at a low angle. The rim of Plato is circular, but from the Earth it appears oblong.

The smooth, flat floor of Plato crater has a relatively low albedo, making it appear dark in comparison to the surrounding rugged terrain. The floor is free of significant impact craters and lacks a central peak. There are a few small craterlets scattered across the floor.

The crater is named for Plato, the great Greek philosopher.

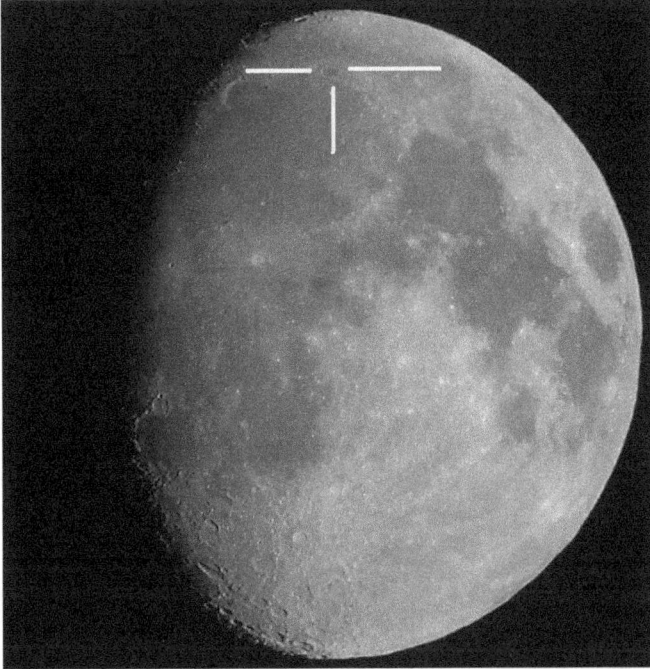

TO LOCATE: Plato is a very distinct feature. Locate Mare Imbrium and look just outside the Mare toward the north.

Archimedes

Archimedes is a large impact crater near the eastern edges of the Mare Imbrium.

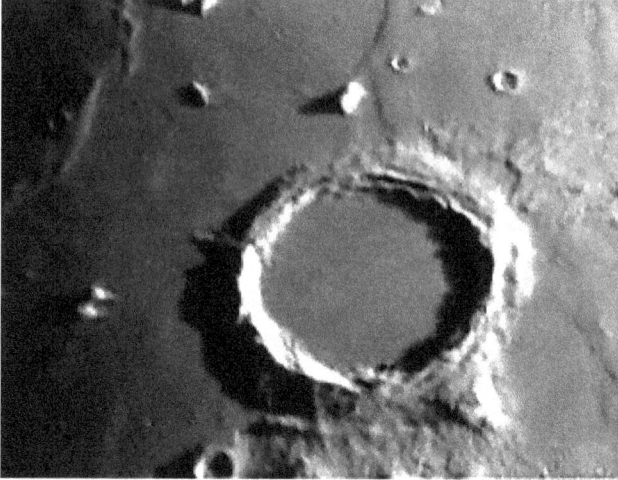

The rim has a significant outer rampart brightened with ejecta but it lacks the ray system associated with younger craters.

The interior of the crater lacks a central peak, and is flooded with maria. It is devoid of significant raised features, although there are a few tiny meteor craters near the rim. Scattered wisps of bright ray material lie across the floor, most likely deposited by the impact that created Autolycus crater.

The crater is named for the Greek philosopher and mathematician.

TO LOCATE: Locate Mare Imbrium. The largest crater inside the Mare Imbrium is Archimedes.

Ptolemaeus

Ptolemaeus is a lunar impact crater. To the south-southeast Ptolemaeus is joined to the rim of Alphonsus crater by a section of rugged, irregular terrain, and these form a prominent chain with Arzachel to the south.

The features of Ptolemaeus are highlighted when the sun is at low angles during the first and last quarter. The crater contours become more difficult to discern during the full moon, when the sun is directly overhead.

The crater has a low, irregular outer rim that is heavily worn and impacted with multiple smaller craters. The crater has no central peak or a ray system. Impact sites of this form are often classified as a "walled-plain", due to their resemblance to the maria.

Ptolemaeus is notable for several "ghost" craters, formed when lava flow covers a pre-existing crater. These leave only a slight rise where the rim existed, and are difficult to detect except at low angles of sunlight.

The crater was named after Claudius Ptolemaeus (c. 90 – c. 168), known in English as Ptolemy, a Greek-speaking geographer, astronomer, and astrologer who lived in the Hellenistic culture of Roman Egypt.

TO LOCATE: Along the terminator of this 8-day moon there are three craters: Ptolemaeus, Alphonsus and Arzachel. Ptolemaeus is the largest of these three, and the one with a smooth floor. In the 10-day moon, these three craters are near the center of the moon.

Alphonsus

Alphonsus is an impact crater located on the lunar highlands on the eastern end of Mare Nubium. It slightly overlaps the Ptolemaeus crater to the north. The surface is broken and irregular along this boundary. The outer walls are slightly distorted. To the northwest is the smaller Alpetragius crater.

The floor is fractured by an elaborate system of rilles and contains three smaller craters surrounded by a symmetric darker halo.

Alphonsus crater was one of the primary alternative landing sites considered for both the Apollo 16 and the Apollo 17 missions. In 1965, the Ranger 9 probe impacted in Alphonsus crater, a short distance to the northeast of the central peak.

The crater is named for Alfonso X (November 23, 1221 – April 4, 1284), a Spanish monarch who ruled as the King of Galicia, Castile and León from 1252 until his death.

locate

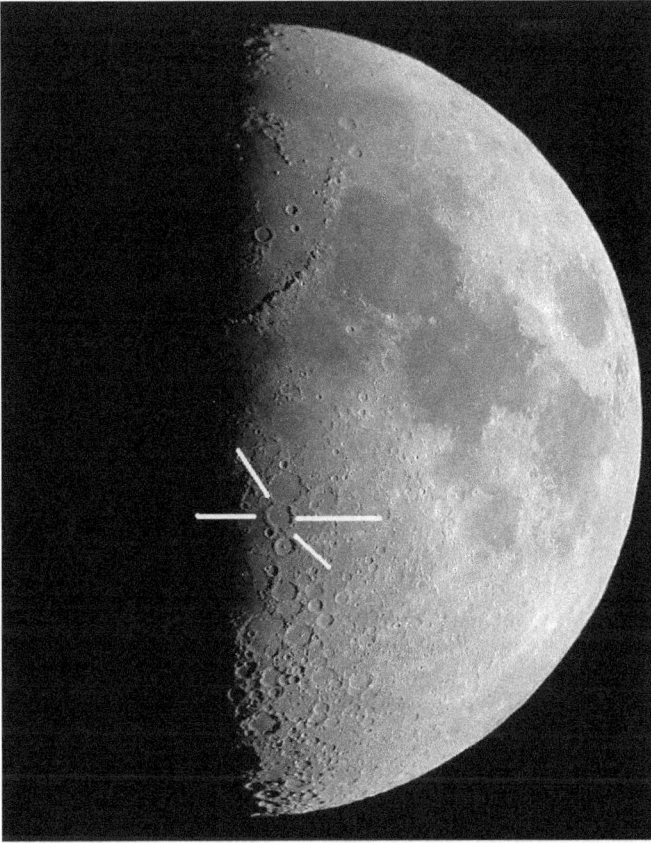

TO LOCATE: Along the terminator of this 8-day moon there are three craters: Ptolemaeus, Alphonsus and Arzachel. Alphonsus is the middle of these three.

Arzachel

Arzachel is an impact crater located in the highlands in the south-central part of the visible Moon, close to the zero meridian (the visible center of the moon). It lies south of Alphonsus crater, and with Ptolemaeus crater further north the three form a prominent line of craters to the east of Mare Nubium.

Arzachel is very clear in its structure and a favorite telescope viewing subject for amateur astronomers. The rim shows little sign of wear and has a detailed terrace structure on the interior, especially on the eastern area of the rim.

There is a rugged central peak that is prominent. The floor is relatively flat, except for some irregularities in the southwestern quadrant of the crater. There is a rille system named the Rimae Arzachel that runs from the northern wall to the southeast rim. A small crater lies prominently in the floor to the east of the central peak, with a pair of smaller craterlets located nearby.

Arzachel was named for Al-Zarqali (rendered as Arzachel in Latin Europe). Al-Zarqali (AD 1028–1087) was a leading Arab mathematician and the foremost astronomer of his time. He excelled at the construction of precision instruments for astronomical use.

TO LOCATE: Along the terminator of this 8-day moon there are three craters: Ptolemaeus, Alphonsus and Arzachel. Arzachel is the smallest and most rugged of these three, and the one with a smooth floor.

Walter

Walter is a lunar impact crater in the western part of the Mare Imbrium.

To the northeast of this crater is the rille named "Rima Diophantus" and the Delisle crater.

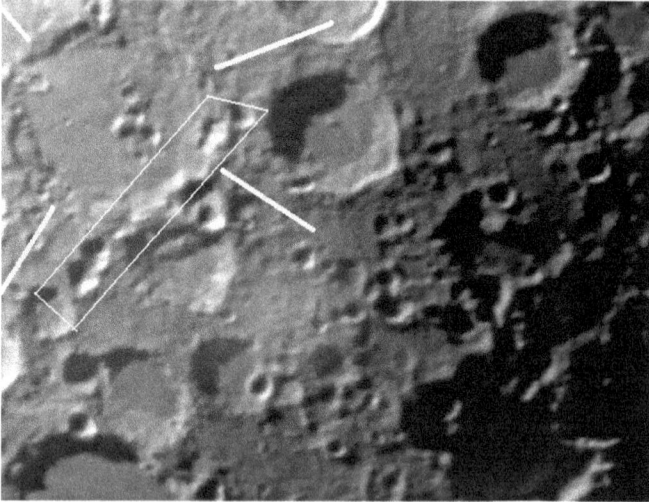

A chain of craters and craterlets line one side of Walter Crater. This line will identify Walter.

Sunrise in Walter can be an interesting time for the astronomer to observe this crater. Small craterlets inside show the tips of their walls while the main crater floor is still dark. Later, light begins to play along the main crater floor.

TO LOCATE: Follow the line of Ptolemaeus and other craters toward Walter. This formation, north of Tycho, has the same low albedo as the surrounding terrain, and is nearly indistinguishable from its surroundings. (Above – the 10-day moon. Below – the 8-day moon.

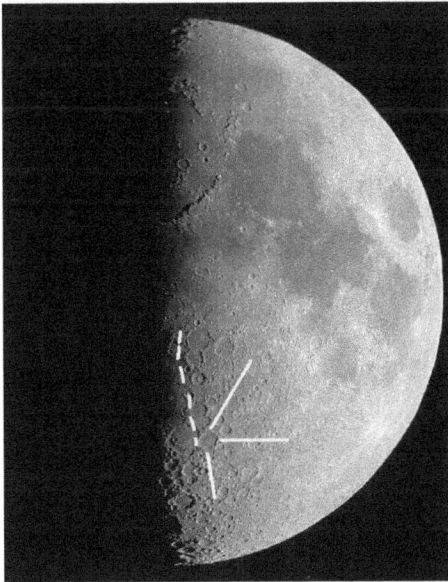

Tycho

Tycho is a very prominent and distinctive impact crater located in the southern lunar highlands. The interior has a high albedo that is prominent when the sun is overhead, and the crater is surrounded by a distinctive ray system forming long spokes reaching hundreds of miles. Sections of these rays can be observed even when Tycho is only illuminated by earthlight. Its inner wall is slumped and terraced. There is a system of central peaks.

Surveyor 7, a robotic spacecraft, landed north of the crater in 1968. Samples recovered from the Apollo 17 mission indicate that Tycho is a relatively young crater, with an estimated age of 108 million years.

The crater has often played a prominent role in science fiction. In the film and book *2001: A Space Odyssey*, Tycho is the location of a mysterious alien monolith which "appears to have been deliberately buried." In *Star Trek: First Contact,* Tycho is the location of a 24th century metropolis.

The crater was named for Tycho Brahe (December 14, 1546 – October 24, 1601). He was a Danish nobleman best known today as an early astronomer, though in his lifetime he was also well known as an astrologer and alchemist.

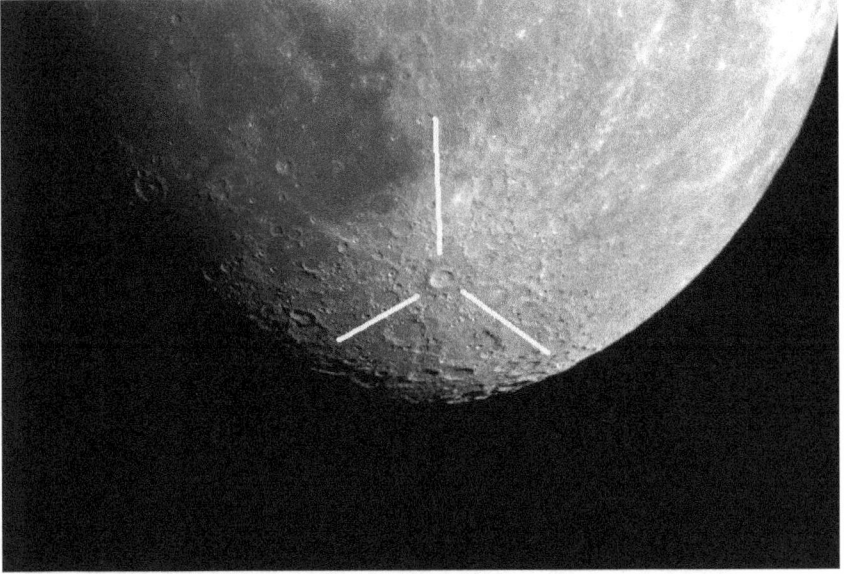

TO LOCATE: Tycho should be an easy formation to find. When the moon is past first quarter, the crater will be prominent because of its ray system. No other crater on the moon has rays that reach such a distance as Tycho.

Maginus

Maginus is an ancient lunar impact crater located in the southern highlands to the southeast of the prominent Tycho crater. (In the photograph below, Maginus is the large crater in the center and Tycho is in the partial crater upper left hand corner).

It is a large formation almost three quarters the diameter of Clavius crater, which lies to the southwest.

The rim of Maginus crater is heavily eroded. The floor is relatively flat, with a pair of low central peaks.

The crater is named for Giovanni Antonio Magini (in Latin, Maginus), an Italian astronomer, astrologer, cartographer, and mathematician.

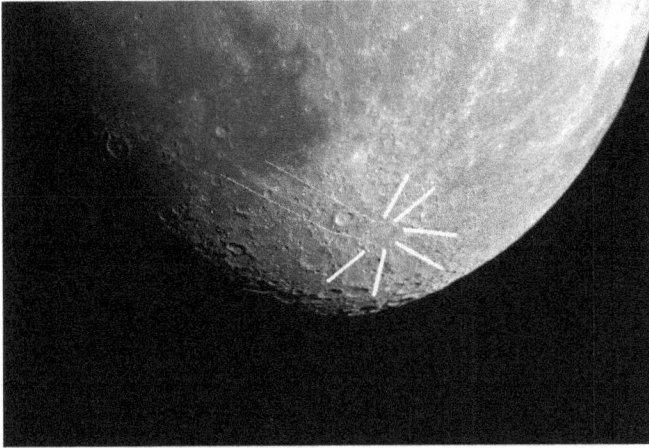

TO LOCATE: Maginus is very close to Tycho. From Tycho, look for a crater that is about a Tycho-size distance from Tycho. If the sun is right, then the rays of Tycho that extend into Mare Imbrium will help draw lines from Imbrium, past Tycho, to Maginus.

Clavius

Clavius is the third largest lunar crater visible from Earth and can be detected with the naked eye. It is located in the rugged southern highlands of the moon, to the south of the prominent Tycho crater.

The crater's floor forms a plain marked by interesting crater impacts. Porter and Rutherford are similar, equal-size craters that intersect the rim of Clavius. The other craters on the floor are Clavius D, C, N, and J, a sequence of diminishing craters that has proved a useful tool for amateur astronomers that want to test the resolution of their small telescopes.

The crater is named for Christopher Clavius, (March 25, 1538 – February 12, 1612), a Jesuit mathematician and astronomer who developed the Gregorian calendar. Clavius doubted the reality of the mountains on the Moon, which is ironic since one of the larges craters on the moon is named for him.

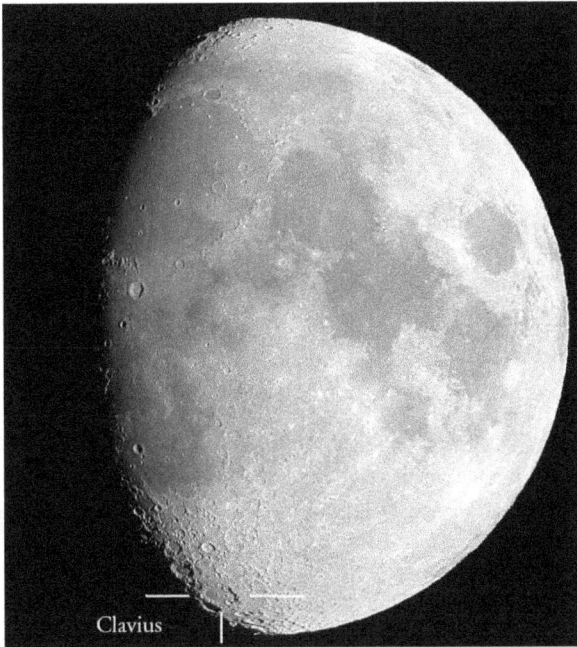

Clavius

TO LOCATE: Clavius appears as a prominent notch in the terminator about 1 to 2 days after the Moon reaches first quarter. It is a prominent and large crater south of the much smaller Tycho.

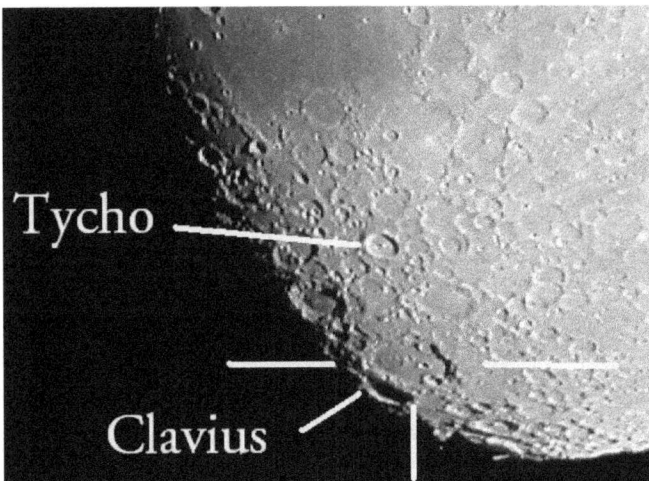

Tycho

Clavius

Copernicus

Copernicus is a prominent lunar impact crater located on the eastern Oceanus Procellarum. It is visible using binoculars.

It is a relatively young crater. Because of its recent formation, the crater floor has not been flooded by lava. The terrain along the bottom is hilly in the southern half while the north is relatively smooth. The central peaks consist of three isolated mountainous rises that are separated from each other by valleys, forming a rough line along an east-west axis.

The crater rays spread across the surrounding maria, overlaying rays from the Aristarchus and Kepler craters. The rays form a nebulous pattern about the crater, making the rays of Copernicus less distinct than the long, linear rays about Tycho crater. In several places the rays lay at glancing angles, instead of forming a true radial dispersal. An extensive pattern of smaller secondary craters can also be observed surrounding Copernicus.

Lunar Orbiter 2 photographed the crater in 1966, creating a detailed image that at the time was named the "Picture of the Century."

The crater was named for Nicolaus Copernicus (1473-1543), an astronomer who provided the first modern formulation of a heliocentric (sun-centered) theory of the solar system.

TO LOCATE: Copernicus is a prominent crater between the areas of Oceanus Procellarum, Mare Nubium, and Mare Imbrium.

Eratosthenes

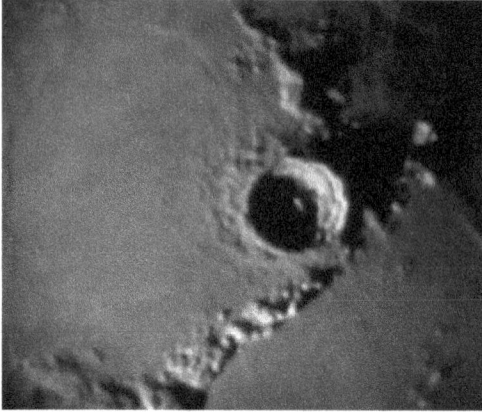

Eratosthenes is a deep lunar impact crater located on the boundary between the Mare Imbrium and Sinus Aestuum. It forms the western terminus of the Montes Apenninus mountain range. The crater has a well-defined circular rim, terraced inner wall, central mountain peaks, an irregular floor, and an outer rampart of ejecta. It lacks a ray system of its own, but is overlaid by rays from the Copernicus crater to the south-west.

At low sun-angles this crater is prominent due to the shadow cast by the rim. When the sun is directly overhead, however, Eratosthenes visually blends into the surroundings, and it becomes more difficult for an observer to locate it. The rays from the Copernicus crater lie across this area, and the higher albedo of these rays serves as a form of camouflage.

The crater is named after Eratosthenes (276 BC - 194 BC), a Hellenistic mathematician, geographer and astronomer. He is noted for devising a system of latitude and longitude, and for being the first known to have calculated the circumference of the Earth.

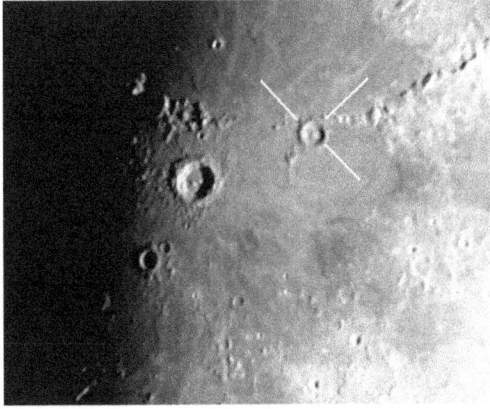

TO LOCATE: Find Mare Imbrium. Follow the mountainous edges toward Copernicus Crater.

Longomontanus

Longomontanus is a lunar impact crater located in the southern highlands to the southwest of Tycho crater. It is of the variety of large lunar formations sometimes called a "walled plain", although it is actually more of a circular depression in the surface. Because of its location, the crater has a distinctly oval in shape due to foreshortening.

To the southeast of Longomontanus is the even larger Clavius crater, and to the east is Maginus crater. North of the rim is the irregular Montanari crater, which in turn is joined at its northern rim by Wilhelm crater.

The wall of Longomontanus is heavily worn by past impacts, and the rim is essentially level with the surrounding terrain. The northern rim especially is impacted with multiple overlapping craterlets. To the east of the rim is a semi-circular ridge that has the appearance of an overlapped crater rim. The crater floor of Longomontanus is relatively flat, with a low cluster of central peaks.

The crater is named for Christian Sørensen Longomontanus (October 4, 1562 – October 8, 1647), a Danish astronomer.

TO LOCATE: With Tycho and Clavius Longomontanus forms a triangle. .

Bullialdus

Bulliadus is an impact crater located in the western part of the Mare Imbrium. The crater is relatively isolated, which highlights its well-formed shape. The inner walls are terraced and contain many signs of landslips. The outer ramparts are covered in a wide ejecta blanket that highlights a radial pattern of low ridges and valleys. In the center of the crater is a formation of several peaks and rises. When the sun is at a high angle, the rim and central mountains appear brighter than the surroundings, and white patches can be viewed on the crater floor.

The crater is named for Ismaël Bulliadus (1605-1694), a French astronomer. He proposed that the force of gravity follows an inverse-square law. In physics, an inverse-square law is any physical law stating that some physical quantity or strength is inversely proportional to the square of the distance from the source of that physical quantity.

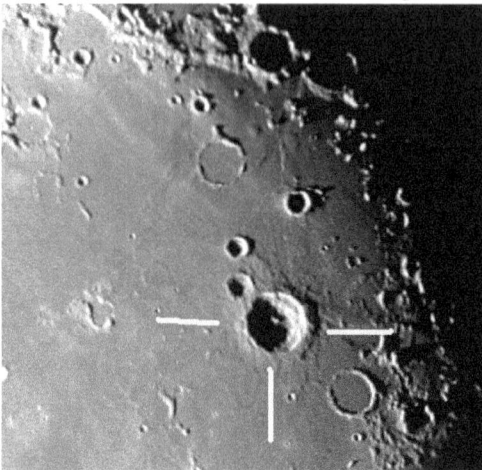

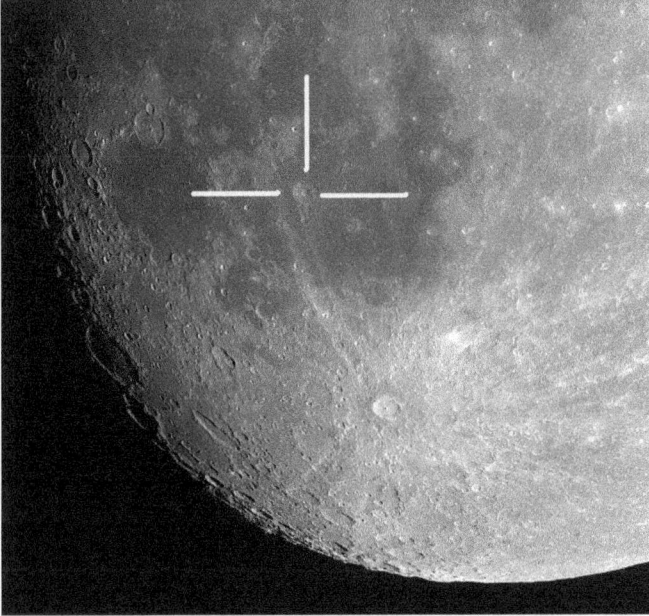

TO LOCATE: Look within Mare Nubium, toward the part of Nubium that is closest to Humorum and Oceanus Procellarum. The crater is somewhat isolated, with very few, but distinctive craters nearby.

Aristarchus

Aristarchus is a prominent lunar impact crater located in the northwest part of the Moon's near side. It is the brightest of the large formations on the lunar surface, with an albedo nearly double that of most lunar features. The feature is bright enough to be visible to the naked eye, and is dazzling in a large telescope. It is also readily identified when most of the lunar surface is illuminated by earth-shine. The crater is located at the southeastern edge of the Aristarchus plateau, an elevated area that contains a number of interesting geological features.

The brightest feature of this crater is the steep central peak. Sections of the interior floor appear relatively level, but Lunar Orbiter photographs revealed a surface covered with many small hills, gouges, and minor cracks and rifts.

The reason for the crater's brightness is that it is a young formation, approximately 450 million years old. This recent formation means the solar wind has not yet had time to darken the excavated material. Aristarchus was formed after the Copernicus crater, but before Tycho crater.

The crater is named for Aristarchus (310 BC-230BC, a Greek astronomer and mathematician. He theorized that the sun, not the Earth, was the center of the universe. Although this heliocentric model of the solar system was rejected, it came 2000 years before Copernicus embraced it. He believed that stars were incredibly distant objects, contrary to the beliefs of his age.

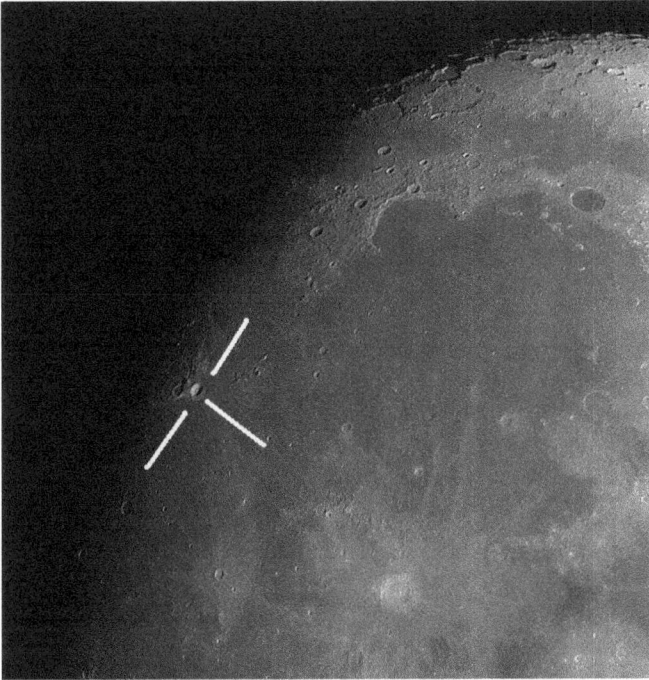

TO LOCATE: Aristarchus should be along the terminator about 2 days after the full moon. Find Copernicus and look into the Mare Imbrium. Aristarchus will be a very bright crater. Even on a Full Moon it should be bright and visible as a feature between Imbrium and Oceanus Procellarum.

Gassendi

Gassendi is a large lunar crater feature located at the northern edge of Mare Humorum. The formation has been inundated by lava during the formation of the mare, so only the rim and the multiple central peaks remain above the surface. A smaller crater 'Gassendi A' is intruding into the northern rim, and joins a rough uplift at the northwest part of the floor. Many observers see a "diamond ring" in this pair of craters.

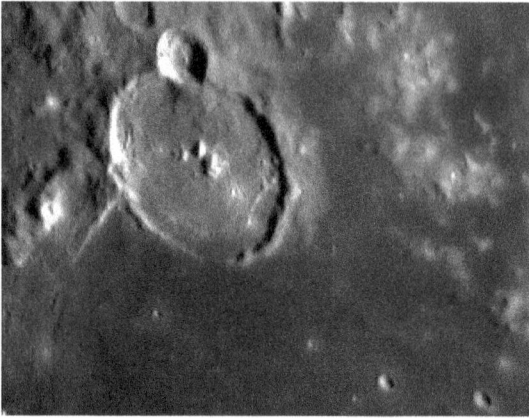

The crater is named for Pierre Gassendi (1592-16), a French philosopher, mathematician and scientist. Gassendi became the first person to observe the transit of a planet across the face of the Sun when he viewed the 1631 transit of Mercury. He watched for the rare transit of Venus in that same year, not knowing that the transit would take place in the evening from his location.

On some older maps the 'Gassendi A' crater was called "Clarkson", after the British amateur astronomer and selenographer Roland L T Clarkson, but this name is not officially recognized by the International Astronomical Union and the name has been removed.

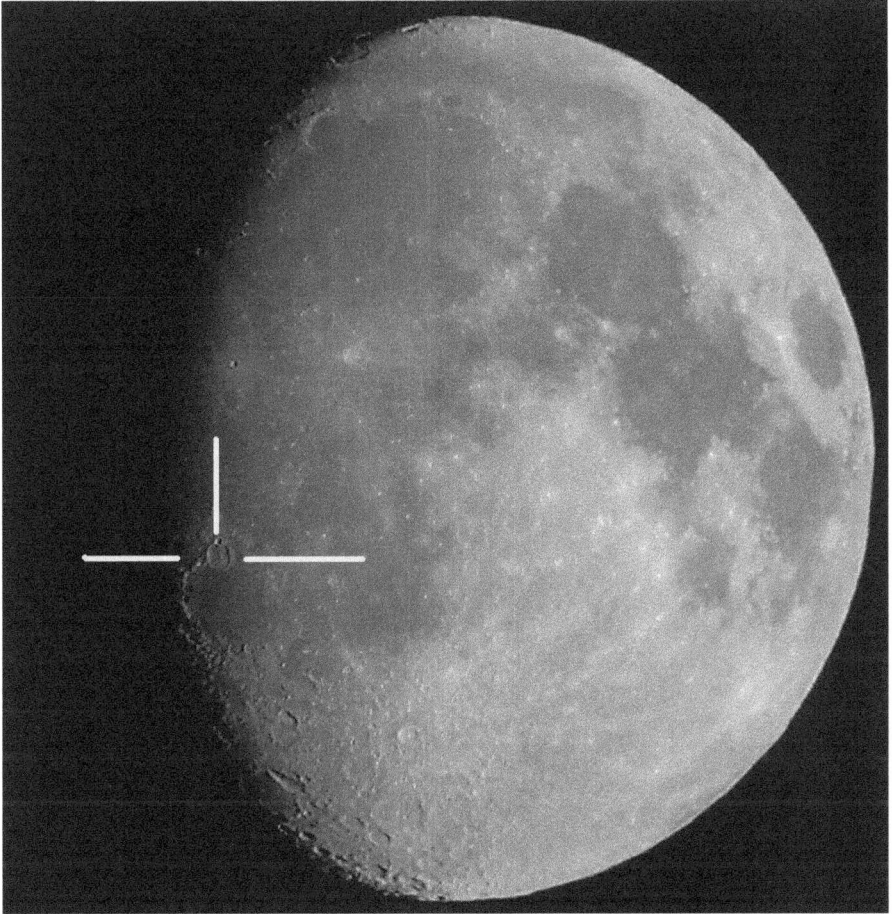

TO LOCATE: Look for Mare Humorum. Gassendi is a large feature on the edge of the Mare.

CRATERS OF THE FOURTEEN DAY OLD MOON

Visible with Binoculars

Kepler

Kepler is a young lunar impact crater located between Oceanus Procellarum to the west and Mare Insularum to the east. Kepler is most notable for the prominent ray system that covers the surrounding maria. The rays overlap the rays from other craters. Kepler has a small rampart of ejecta surrounding the exterior of its high rim. The interior walls of Kepler are slumped and slightly terraced, descending to an uneven floor.

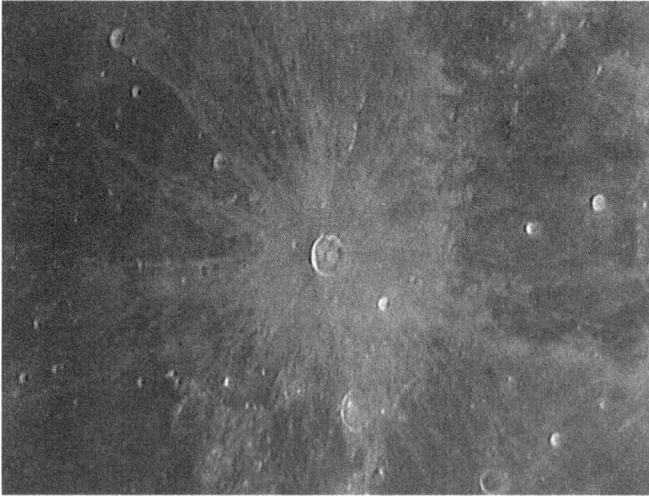

The crater was named for Johannes Kepler (1571-1630), a mathematician and astronomer. He is best known for his laws of planetary motion. The name was given by Giovanni Riccioli, who noticed one of the rays from Tycho crater extended across Oceanus Procellarum and intersected Kepler crater. Riccioli saw this as a tribute to how Kepler used the observations of Tycho Brahe while devising his three laws of planetary motion.

TO LOCATE: Look for Copernicus and move into the edge of the Oceanus Procellarum.

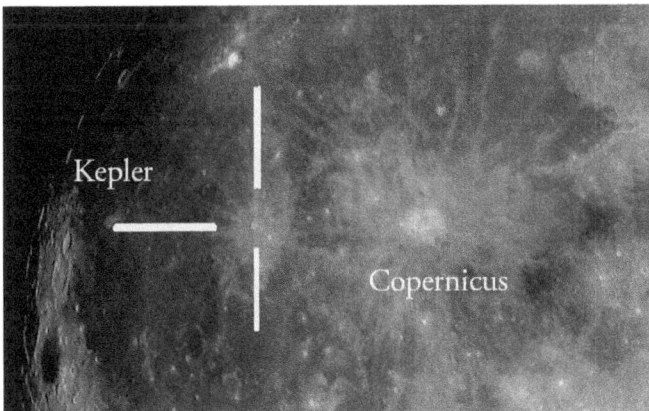

Grimaldi

Grimaldi is a large basin located near the western limb of the Moon, located on the southwest of the Oceanus Procellarum, and southeast of Riccioli crater.

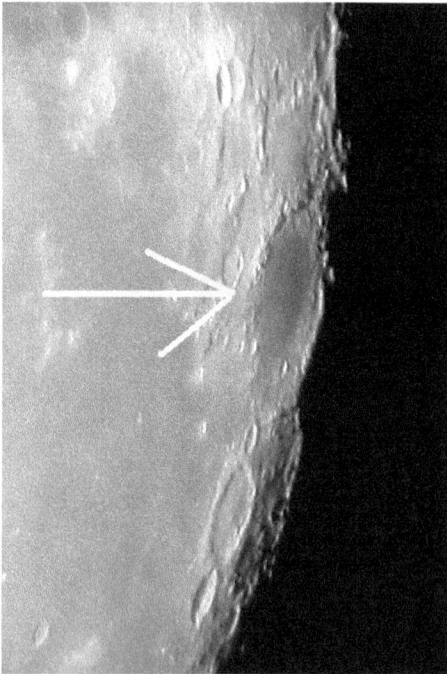

The inner wall of Grimaldi has been so heavily worn and eroded by subsequent impacts that it forms a low, irregular ring of hills, ridges and peaks, rather than a typical crater rim. The floor is the most notable feature of this crater, forming a flat, smooth and featureless surface with a particularly low albedo, which contrasts the brighter surroundings. This contrast should make the crater easy for observers to find.

The crater is named for Francesco Maria Grimaldi (1618-1663), an Italian mathematician and physicist. In astronomy, he built and used instruments to measure geological features on the Moon.

TO LOCATE: Follow Copernicus, then Kepler toward the limb, where you will see a dark circle, which is Grimaldi.

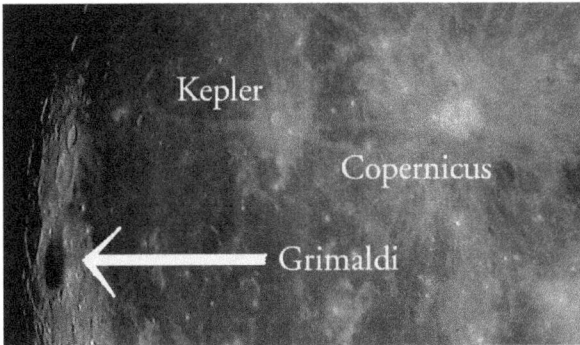

Kepler

Copernicus

Grimaldi

LUNAR FEATURES

Visible with a Telescope

Sinus Aestuum

This mare-like area forms a northeastern extension to the Mare Insularum. It is a level, nearly featureless surface of low albedo basaltic-lava, marked by a few small impacts and some ridges. The eastern border is formed by an area of irregular terrain that divides the bay from the Mare Vaporum to the east. The Montes Apenninus range is to the north. The prominent Eratosthenes crater is also toward the north. Along the western side is the flooded Stadius crater and the Mare Insularum to the southwest.

The name of this plain is Latin for "Bay of Billows".

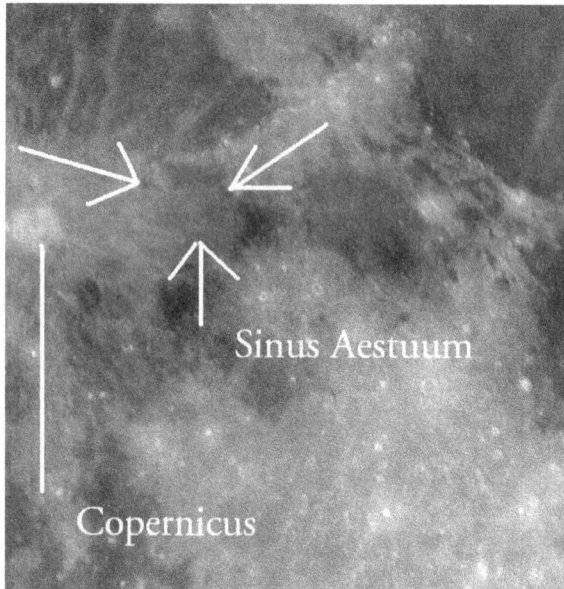

Lacus Mortis

Lacus Mortis, Latin for "Lake of Death", is a plain of basaltic-lava flows in the northeastern part of the Moon. It lies just to the south of the elongated Mare Frigoris, being separated by a slender arm of rugged ground. To the south is the Lacus Somniorum, separated from this mare by the Plana-Mason crater pair.

The Bürg crater is located prominently just to the east of the mid-point of this feature. The western part of the Lacus Mortis contains an extensive system of criss-crossing rilles collectively designated Rimae Bürg.

Palus Putredinis

Palus Putredinis (Latin for "Marsh of Decay") is an area of the lunar surface that stretches from Archimedes crater southeast toward the rugged Montes Apenninus range. The region is a nearly level, lava-flooded plain. It is bounded by Autolycus crater to the north and the foothills of the Montes Archimedes to the west.

In the southern part of this area is a rille system designated Rimae Archimedes. To the south is a prominent linear rille named Rima Bradley, and to the east is the Rima Hadley and the Rimae Fresnel. Just to the northwest of the Palus Putredinis mid-point is the nearly submerged Spurr crater.

Promontorium Laplace

Promontorium Laplace is the northeastern tip of the Jura Mountain range that cusps the bay of Sinus Iridum which itself is a northwestern extension to Mare Imbrium. These mountains rise high above the smooth floor of Iridum.

It is named for Pierre Simon marquis de Laplace, an 18th Century French astronomer, mathematician and physicist.

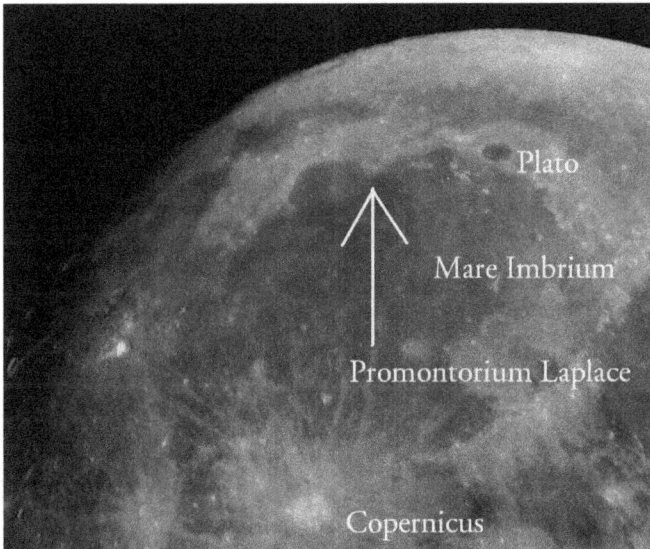

Promontorium Heraclides

Promontorium Heraclides it was named after Heraclides Ponticus (390 - 310 BC) a student of Plato who believed the Earth rotated on its axis.

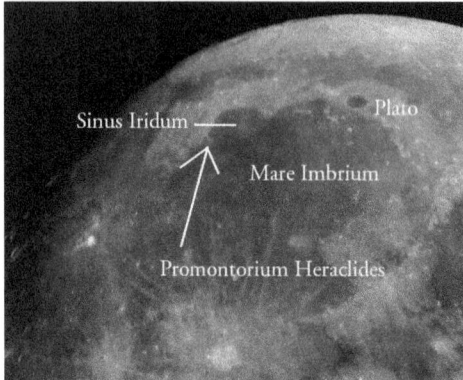

According to an interesting note in the Hatfield atlas, one might keep on the look out for the "Moon Maiden" when Promontorium Heraclides is illuminated by the right amount of sunlight! Under the right lighting conditions, when the moon is less than 11 days old, this formation transforms itself into what Cassini called the "Moon Maiden". Given proper illumination and not the best seeing (meaning low resolution) the maiden appears out of the lunar surface.

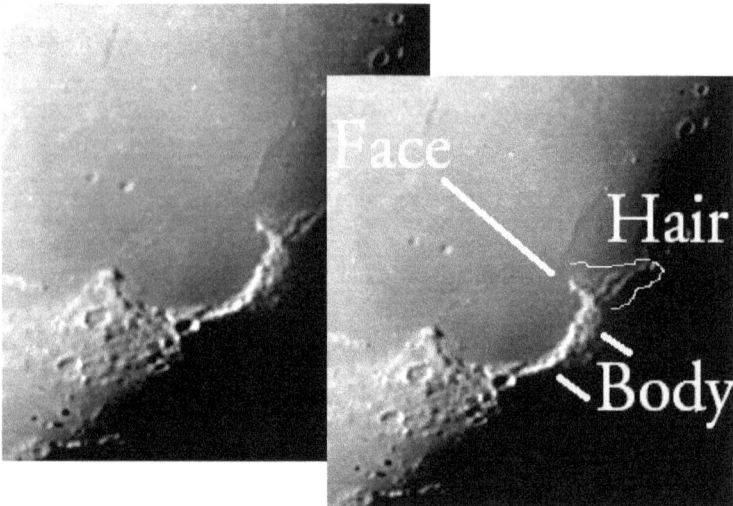

Promontorium Agarum

The bright point on the shoreline of Mare Crisium is Promontorium Agarum with the shallow crater Condorcet to its east. Look along the shore of the mare for a mountain to the south known as Mons Usov. Just to its north Luna 24 landed and directly to its west are the remains of Luna 15.

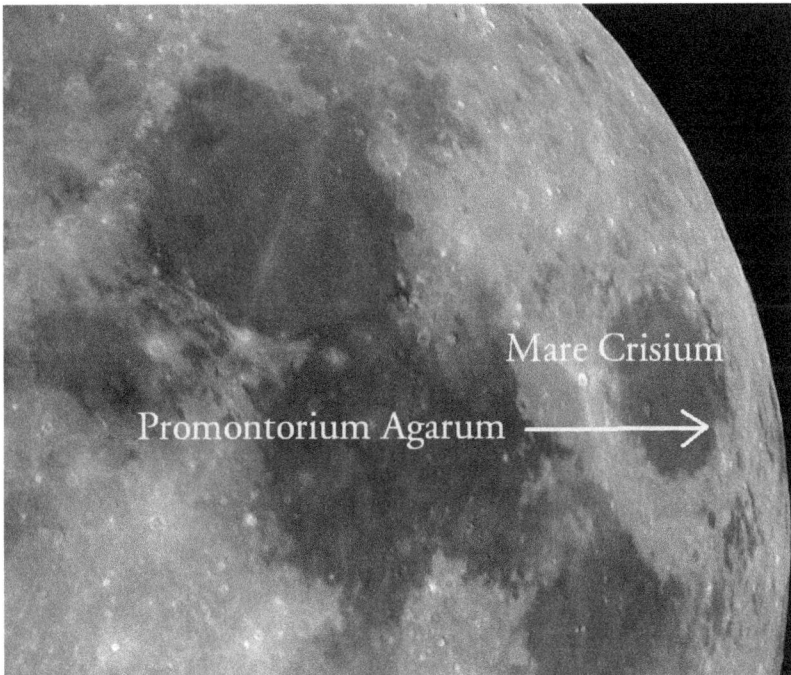

Montes Alpes

Montes Alpes is a mountain range in the northern part of the Moon's near side. It was named for the Alps in Europe. This range forms the northeastern border of the Mare Imbrium lunar mare. To the west of the range is the level and nearly featureless mare. The eastern face is a more rugged area with a higher albedo. The range begins about one crater diameter northwest of Cassini crater, at the Promontorium Agassiz, then stretches about 30 miles to the northwest then continues to the eastern rim of Plato crater.

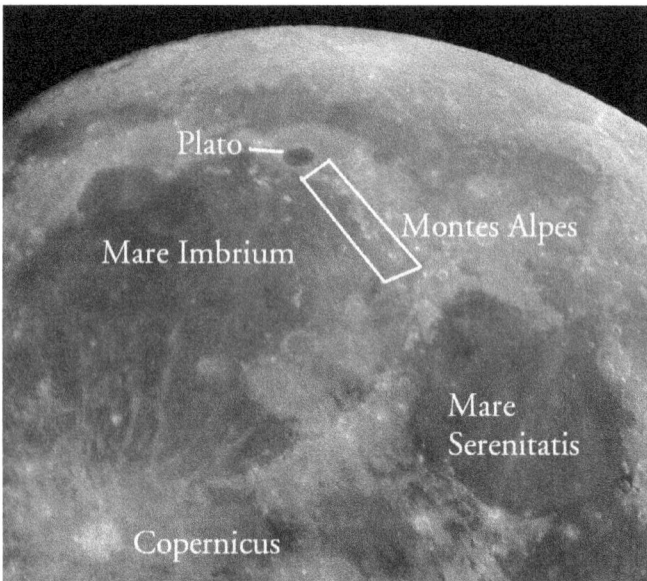

Montes Apenninus

Montes Apenninus are a 370 mile-long rugged mountain range on the northern part of the Moon's near side. They named for the Apennine Mountains in Italy.

The range forms the southeastern border of the large Mare Imbrium lunar mare. It begins just to the west of the prominent Eratosthenes crater.

This range contains several mountains that have received names, listed below ranging from west to northeast: Mons Wolff, Mons Ampère, Mons Huygens, Mons Bradley, Mons Hadley Delta, and Mons Hadley. The last two are perhaps familiar to many because they form the valley where the Apollo 15 mission made its landing

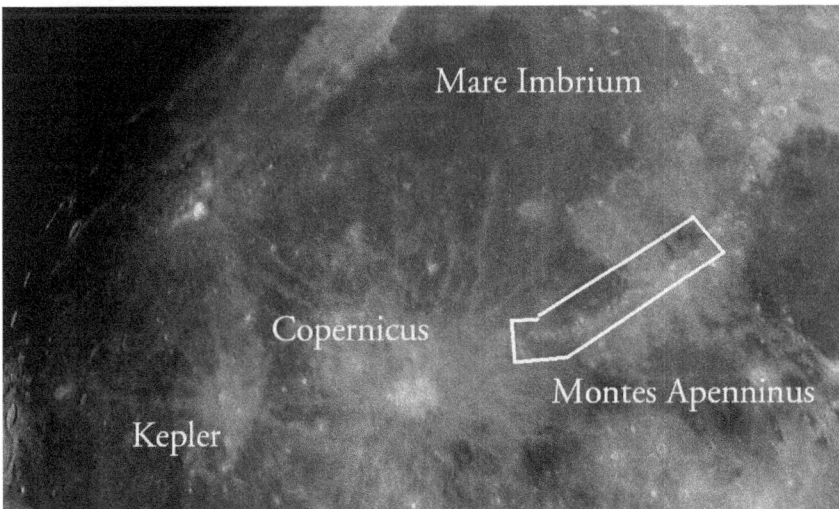

Mons Hadley

Mons Hadley is an example of a massif on the Moon,[1] located at the northern area of the Montes Apenninus.

To the south of Mons Hadley is a valley that served as the landing site for the Apollo 15 expedition.

Apollo 15 astronaut, with Mons Hadley in the background.

---

[1] In geology, a massif is a section of the Earth's crust that is demarcated by faults or flexures. In the movement of the crust, a massif tends to retain its internal structure while being displaced as a whole.

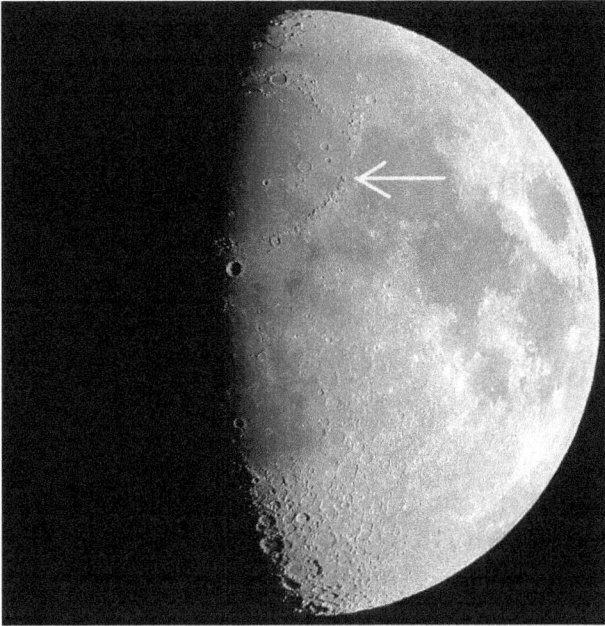

TO LOCATE: To locate, look for three large craters in Mare Imbrium and use them to point to Mons Hadley.

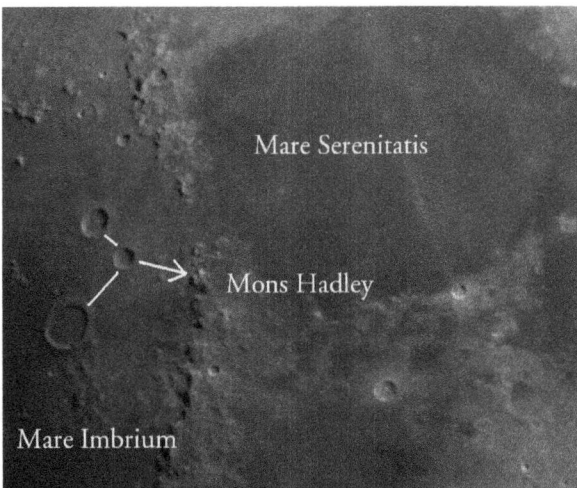

Mons Piton

Mons Piton is an isolated lunar mountain in the eastern part of the Mare Imbrium, to the north-northwest of Aristillus crater. Due east of Mons Piton is Cassini crater, and to the west-northwest lies the Piazzi crater.

It is slightly elongated along toward the northwest, with ridge lines to the south, northwest, and west.

Because it is an isolated formation on the lunar mare, this peak can form prominent shadows when illuminated by oblique sunlight during the lunar dawn or dusk.

Mons Piton was named for a peak on Tenerife Island, the largest of the seven Canary Islands off the coast of Africa.

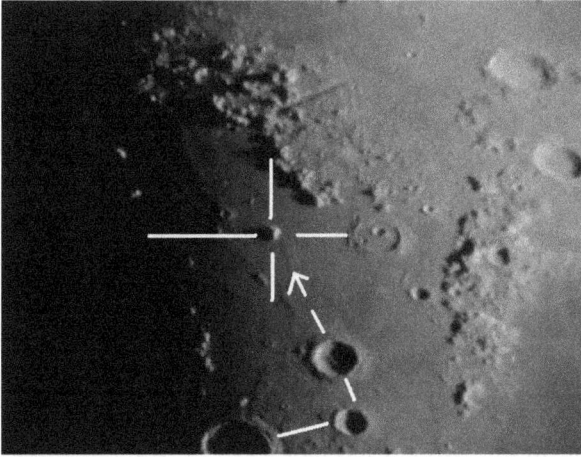

TO LOCATE: The same three craters that make
an arrow pointing to Mons Hadley also form an "L"
that leads to Mons Piton.

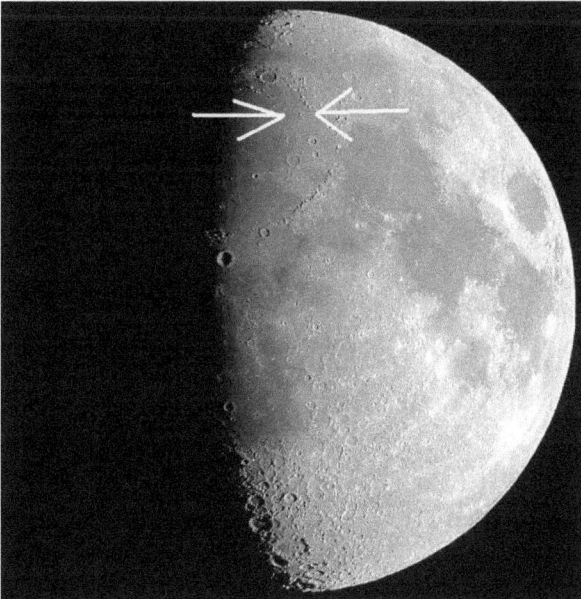

Mons Pico

Mons Pico is a solitary lunar mountain in the northern part of the Mare Imbrium basin, and south of the prominent Plato crater. This peak forms part of the surviving inner ring of the Imbrium basin. This ring continues to the northwest and with the Montes Teneriffe and Montes Recti ranges.

Because it is in an isolated location on the lunar mare, Mons Pico can form prominent shadows when illuminated by oblique sunlight.

A smaller peak to the south of Mons Pico is designated Pico B. This region of the mare is notable for a number of wrinkle ridges.

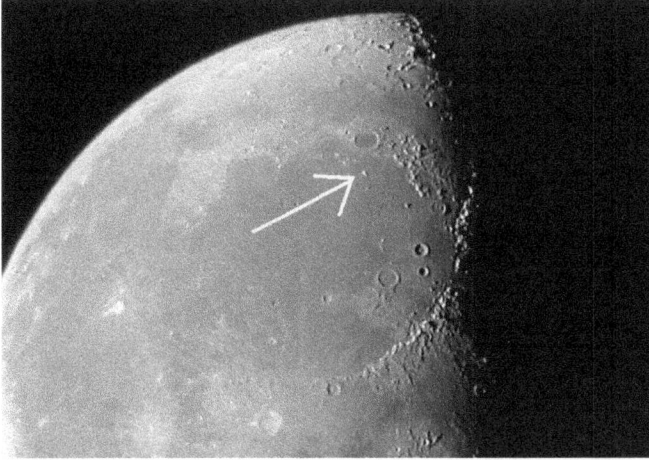

TO LOCATE: Look near Plato. Find Vallis Alpes. Follow that feature to two bright mountains. Pico B seems aligned with Vallis Alpes, while Mons Pico is perpendicular to the "line" formed by Vallis Alpes.

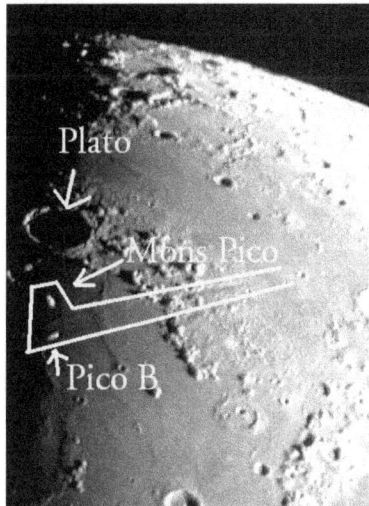

Rupes Altai

Rupes Altai is a 265 mile long cliff on the lunar surface that forms the southwestern rim of the Nectaris impact basin.

The southeastern end of the cliff terminates along the western edge of the Piccolomini crater. It then arcs irregularly toward the north. The northern end of the arc is an irregular region with no clearly defined terminus, where it brackets the prominent Theophilus, Cyrillus, and Catharina craters.

This feature is difficult to locate during the full moon when the sunlight is nearly overhead. It is best viewed when the terminator is nearby and the sunlight is arriving at a low angle.

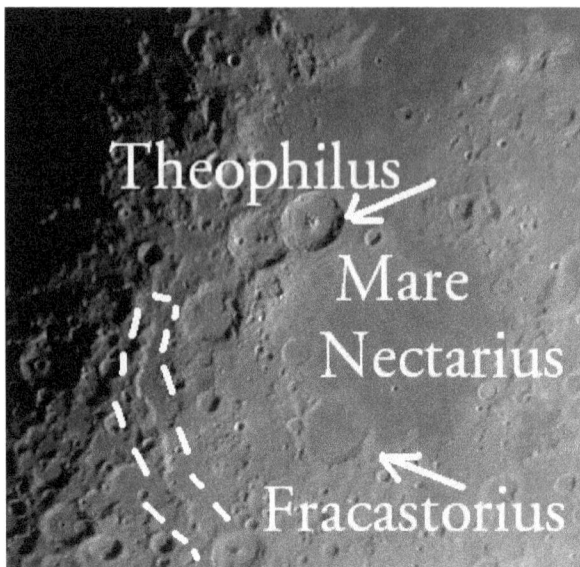

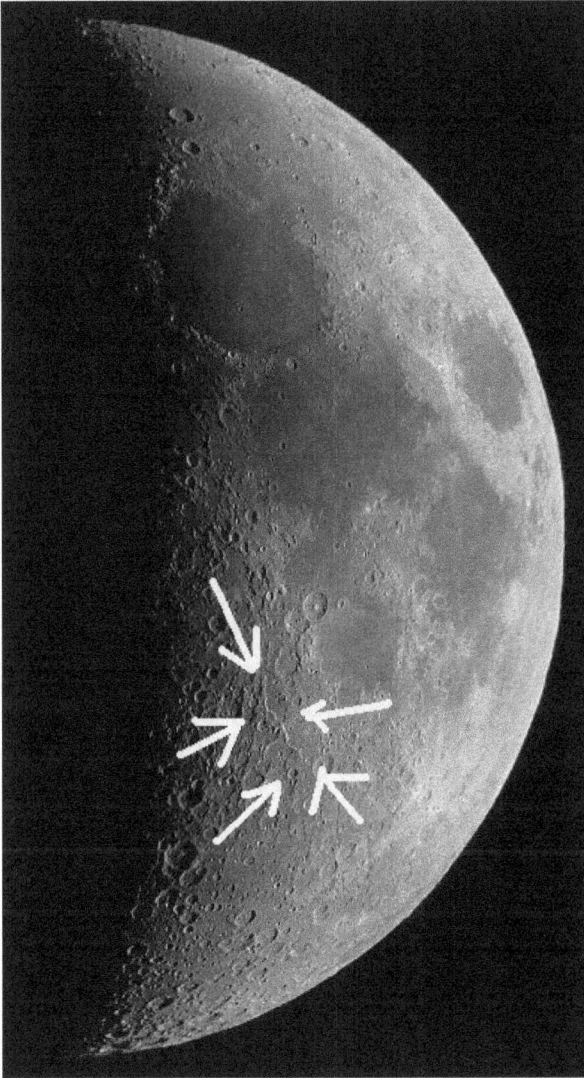

TO LOCATE: Find Mare Nectaris. Find the three prominent craters – Theophilus, Cyrillus and Catharina. Follow the line of those three craters, starting at Theophilus, to the Rupes Altai.

Rima Hyginus

Rima Hyginus is a large rille at the east end of the Sinus Medii. The southeast-northwest oriented feature crosses over the crater Hyginus, and then takes a west-east orientation.
.

Rima Hyginus is a distinctive and prominent feature in an otherwise flat surface. Smaller craterlets can also be discerned along the length of this rille, possibly caused by a collapse of an underlying structure.

The Rima Hyginus, or Hyginus Rille, is a favorite target for amateur telescopes. Nearby are the Ariadaeus Rille and the Triesnecker Rilles, making this part of the lunar surface unusually rich in these somewhat mysterious clefts.

It was named for Caius Julius Hyginus, a 2nd Century BC Greek astronomer.

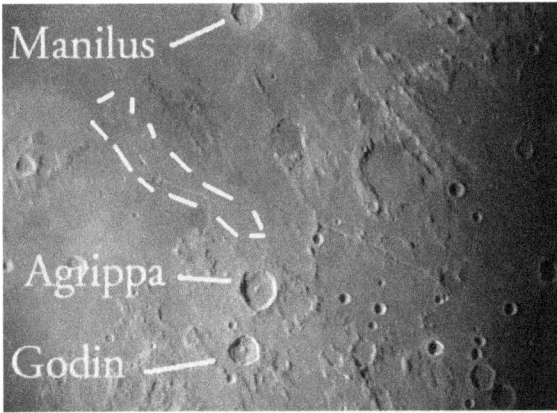

TO LOCATE: Locate Sinus Medii. Locate the craters Manilus (#1 below, Agrippa (2) and Godin (3). Rima Hyginus is located between Agrippa and Manilus.

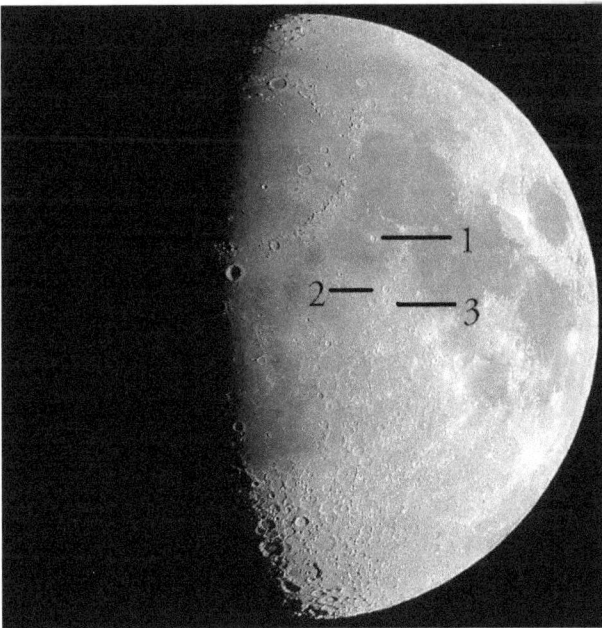

Vallis Schroteri

The Vallis Schroteri region is an area of never ending awe. This part of Oceanus Procellarum is rich in volcanic structures. Considered as a possible landing site for the Apollo missions, it lost out to Hadley Rille as Apollo 15's landing site. Among the many interesting features in this area is the bright young crater, Aristarchus. It almost looks like someone put a dab of white paint on a crater. Aristarchus is even visible on the night side of the moon during Earthshine.

Telescopes of all sizes can pick out Vallis Schroteri, the largest sinuous valley on the moon. At the beginning of the snakelike valley is the cobra head feature, a widening just north of a tiny crater. The valley meanders in a "U" shape for about 100 miles from the crater Herodotus to the south.

TO LOCATE: Look in the Oceanus Procellarum. Crater Aristarchus is the bright crater in the Oceanus. Focus your telescope on that area to see the snake-like Valles Schroteri

Vallis Alpes

Vallis Alpes (Latin for"Alpine Valley") is a spectacular 103 mile valley feature that bisects the Montes Alpes range. It is narrow at both ends and widens to a maximum width of about 6 miles along the middle stretch. The valley floor is a flat, lava-flooded surface that is bisected by a slender, cleft-like rille. (This cleft is a challenging target for telescope observation from the Earth.)

The sides of the valley rise from the floor to the surrounding highland terrain, a blocky, irregular surface. The northern face of the valley is not as straight as the southern side.

This valley was discovered in 1727 by Francesco Bianchini.

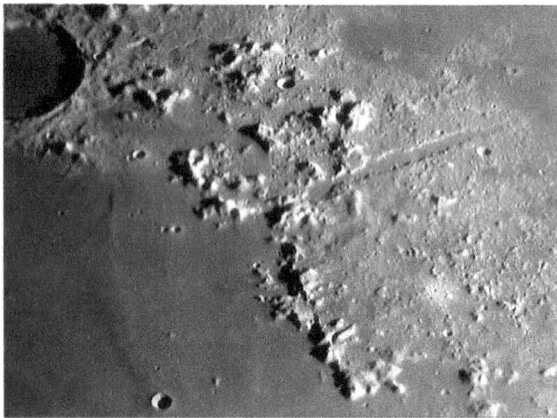

The gash above the center of this photo shows Vallis Alpes.

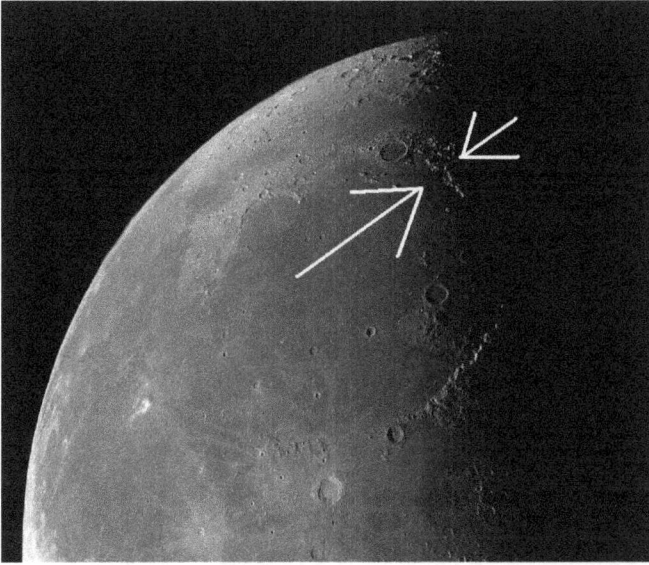

TO LOCATE: Find Mare Imbrium. Look near Plato crater. Look along the border of Mare Imbrium from Plato, on the opposite side of Plato that one would find Sinus Iridum.

Rupes Recta
"The Straight Wall"

Rupes Recta is commonly called "the Straight Wall," but it is more accurate to refer to it as the "Straight Fault." The proper name, "Rupes Recta," actually means "Straight Fault."

Rupes Recta is a linear fault on the Moon, in the southeastern part of the Mare Nubium. It is not a steep scarp but a moderate slope of about 7 degrees. When illuminated by the rising Sun from the East (less than a day after First Quarter) it casts a striking shadow. It is a very popular target for amateur astronomers. With a remnant of a crater at one end of the fault, it gives the appearance of a sword.

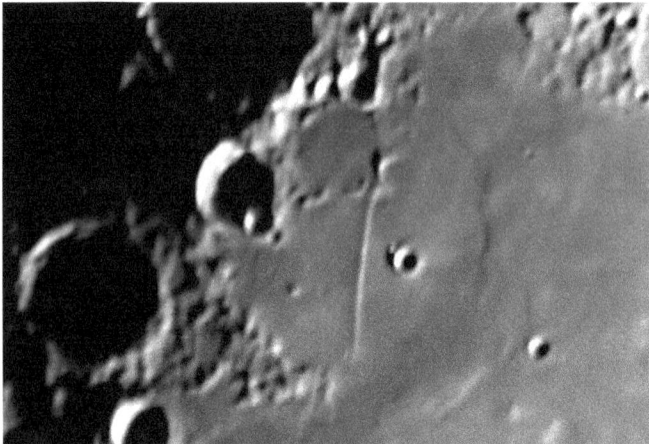

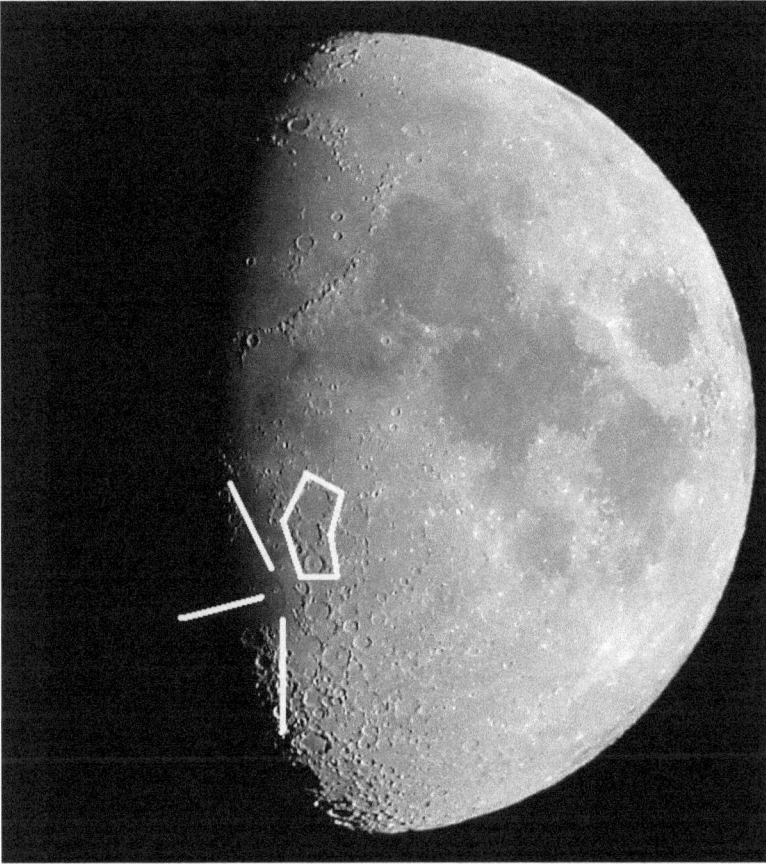

TO LOCATE: Locate the three prominent craters: Ptolemaeus, Alphonsus and Arzachel. From the smaller of these three – Arzachel – look into Mare Nubium.

CRATERS OF THE FOUR DAY OLD MOON

Visible with a Telescope

Picard

Picard is a lunar impact crater that lies in the western part of the Mare Crisium. To the west is the almost completely flooded Yerkes crater. The tiny Curtis crater is east of Picard.

The crater rim of Picard is well-defined and shows little sign of wear, having a sharp-edged appearance. It has a low hill at the center.

The crater is named for Jean-Felix Picard (July 21, 1620 – July 12, 1682), a French astronomer and priest. He was the first person to measure the size of the Earth to a reasonable degree of accuracy.

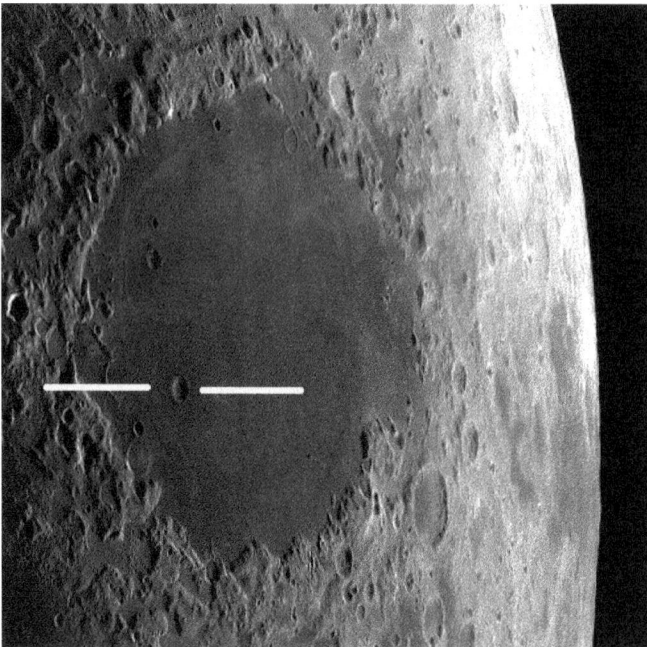

TO LOCATE: Look in Mare Crisium on the side that is away from the nearby lunar limb. Look for the large and solitary.

Furnerius

Furnerius is a lunar crater located in the southeast part of the Moon, in the area close to the southwestern limb. Because of its location, the crater appears oval in shape due to foreshortening but is actually nearly circular.

The rim of Furnerius is worn and battered, with multiple impacts along its length and notches along the base. Much of the wall now rises only slightly above the surrounding terrain, with the lowest sections to the north and south.

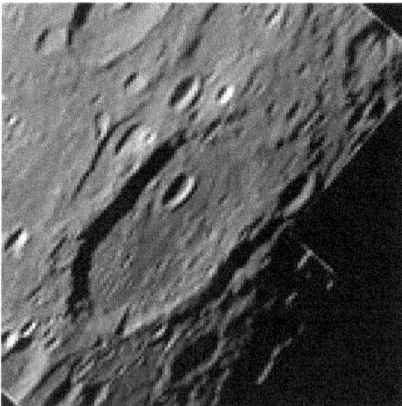

The interior floor is marked by fourteen notable craters, the most notable being 'Furnerius B' in the northern half which has a central rise. In the northeast part of the floor is a rille designated Rima Furnelius.

The Japanese satellite, Hiten, crash-landed in the vicinity of this crater in 1993.

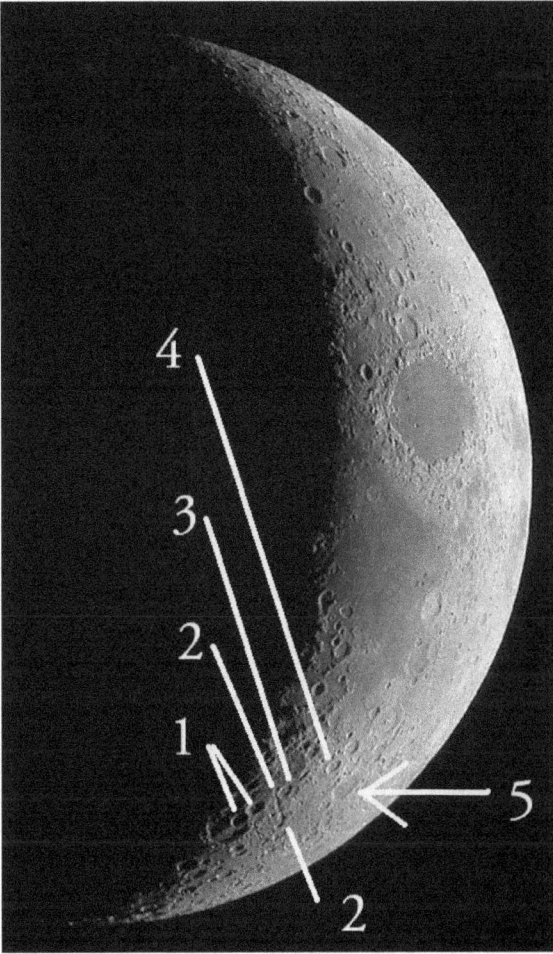

TO LOCATE: Look along the southern area of the 4 day moon. Find a pair of craters (1), then a fault or line (2). Just over the "line" is another crater (3). Jump a distance away to a similar crater (4). From that crater move toward the limb and slightly south to a larger crater (5). That is Furnerius.

Petavius Wall

Petavius is a large impact crater located to the southeast of the Mare Fecunditatis, near the southeastern lunar limb. The crater Petavius appears oblong when viewed from the Earth due to foreshortening. The outer wall of Petavius crater is unusually wide in proportion to the diameter, and displays a double-rim along the south and west sides.

The Reverend T. W. Webb described Petavius Crater as, "one of the finest spots in the Moon: its grand double rampart, on east side nearly 11,000 ft. High, its terraces, and convex interior with central hill and cleft, compose a magnificent landscape in the lunar morning or evening, entirely vanishing beneath a Sun risen but halfway to the meridian."

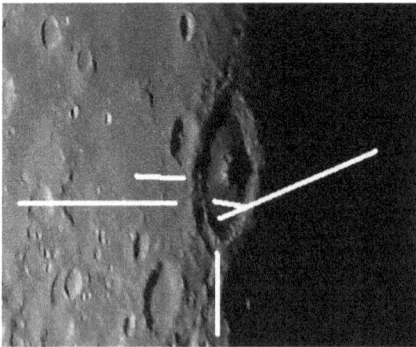

The most favorable time for viewing this feature through a telescope is when the Moon is only three days old. By the fourth day the crater is nearly devoid of shadow.

The feature is named for Denis Petau, also known as Dionysius Petavius (1583-1652), a French Jesuit theologian.

TO LOCATE: Look along the southern edge of Mare Fecunditatis. Just outside the maria is the crater, Petavius. To find the wall, you need about a 3 day moon. The larger photograph above is a 4 day moon. The small insert is a 3 day moon. Notice the difference in the wall.

Messier/Messier A

Messier is a relatively young impact crater located on the Mare Fecunditatis. The crater has a discernable oblong shape that is not caused by foreshortening. The longer dimension is oriented in an east-west direction.

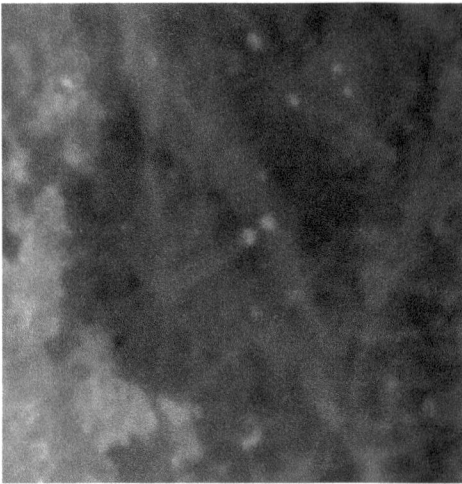

What makes these two craters so interesting to the observer is the distinct shape. This pair of lunar features has a bright double ray system reaching out from the two craters to the West, giving an appearance of a comet with a tail. The interior of craters Messier and Messier A have a higher albedo than the surrounding maria, adding to the appearance of the comet image.

It is theorized that Messier crater was formed by an impact at a very low angle, and that 'Messier A' could have formed following a rebound by the impacting body. The low angle of impact may also explain the asymmetrical ray system.

The feature is named for Charles Messier, who is familiar to amateur astronomers for his catalog of deep sky objects.

TO LOCATE: Look at the interior of Mare Fecunditatis. The comet tail shape should be easy to find.

Proclus

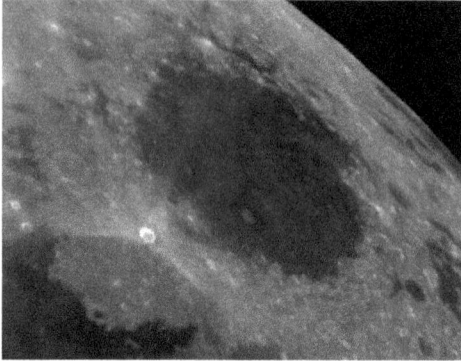

Proclus is a young impact crater that is west of Mare Crisium. The rim of Proclus crater is polygonal in shape, having the shape of a pentagon, and does not rise very far above the surrounding terrain. It has a high albedo, being second only to Aristarchus crater in brightness. The interior wall displays some slumping, and the floor is uneven with a few small rises from slump blocks.

The crater has a notable ray system. The rays display an asymmetry of form, with the most prominent being rays to the northwest, north-northeast, and northeast. There is an arc with no ejecta to the southwest. These features suggest an oblique impact at a low angle.

The crater is named for Proclus (412-485), a Greek Neoplatonist philosopher, and one of the last major Greek philosophers.

TO LOCATE: Look for the Mare Crisium, then find the bright crater on the outside of the maria, between Crisium and the 4 day old moon's terminator.

158

Fabricius

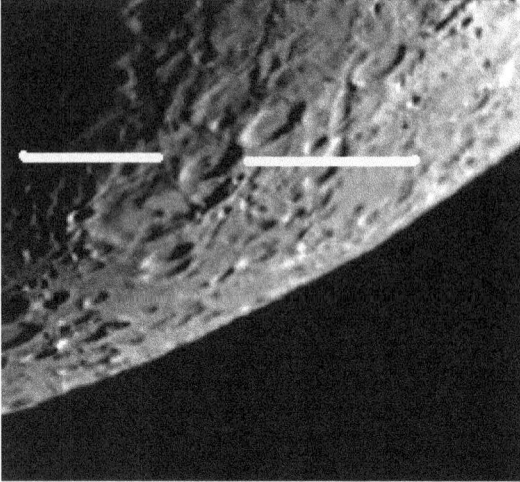

Fabricus is an impact crater that is located within the northeast part of the Janssen walled plain. Attached to the north-northwest rim is the slightly larger Metius crater. The crater has multiple central peaks.

This feature is named for David Fabricius (March 9, 1564 - May 7, 1617), a German theologian. As was common for churchmen of the day, he dabbled in science: his particular interest was astronomy and he made two major discoveries in those early days of telescopic astronomy.

Fabricius discovered the first known periodic variable star (as opposed to cataclysmic variables, such as novas and supernovas), Mira, in August of 1596. When he saw Mira brighten again in 1609, however, it became clear that a new kind of object had been discovered in the sky. Two years later, he and his son used their telescope to observe the Sun. They noted the existence of sunspots, the first confirmed instance of their observation. Little else is known about Fabricius except that he was murdered by a parishioner.

TO LOCATE: Along the terminator border of a 4-day moon, look along the southern terminator. Fabricius and Metius form a figure "8." Fabricius is the southern-most of these two.

CRATERS OF THE SEVEN DAY OLD MOON

Visible with a Telescope

Plinius

Plinius is a prominent impact crater lies on the border between Mare Serenitatis in the north and Mare Tranquillitatis in the south.

The sharp rim of the Plinius crater is slightly oval in form, with a terraced inner wall, and an irregular outer rampart. It does not have a ray system. The crater's floor is hilly, and in the middle is an irregular central peak that has the appearance of a double crater formation under certain angles of illumination.

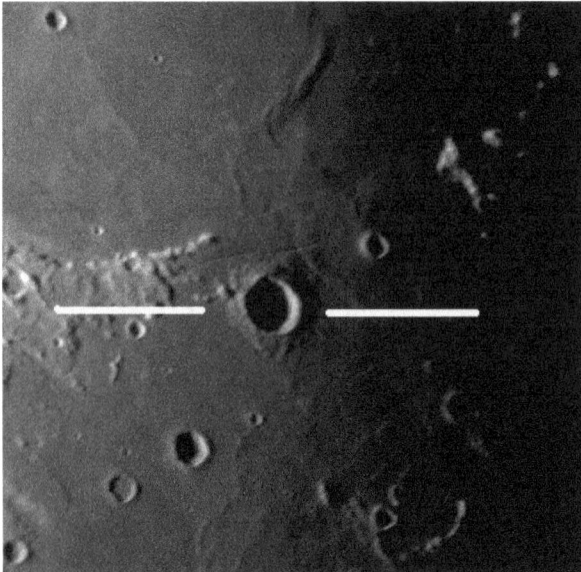

This feature is named for Gaius Plinius Secundus, (23–79) better known as Pliny the Elder. He was an ancient author and Natural philosopher of some importance who wrote *Naturalis Historia*.

TO LOCATE: Look for the Mare Tranquillitatis and Mare Serenitatis. Look at the area between the two maria. Plinius should stand out as a prominent crater between the two.

Mitchell

Mitchell is a crater attached to the eastern rim of the larger and more prominent Aristoteles crater. The floor of Mitchell is rough and irregular, with a low central rise. There is a slight notch in the southern rim, and the western wall has been completely absorbed by the Aristoteles rim.

The feature is named for Maria Mitchell (August 11, 1818 – June 28, 1889), an American astronomer. Using a telescope, she discovered "Miss Mitchell's Comet" (Comet 1847 VI, modern designation is C/1847 T1) in the autumn of 1847. This gave her worldwide fame, as she was only the second woman to discover a comet (the first being Caroline Herschel).

She was the first woman member of the American Academy of Arts and Sciences and the American Association for the Advancement of Science. She became professor of astronomy at Vassar College in 1865. She was also named as Director of the Vassar College Observatory. After teaching there for some time, she learned that despite her reputation and experience, her salary was less than that of many younger male professors. She insisted on, and received, a salary increase.

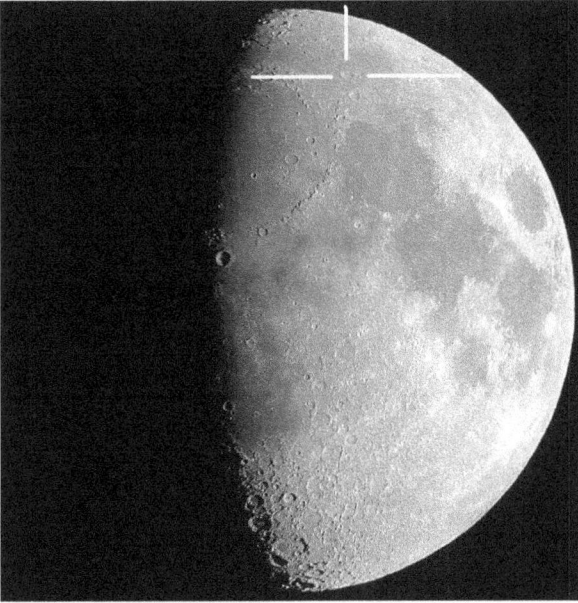

TO LOCATE: Look in the Frigoris. Mitchell is in this area. Find the large crater of Aristoteles. Mitchell is connected to the rim of Aristoteles.

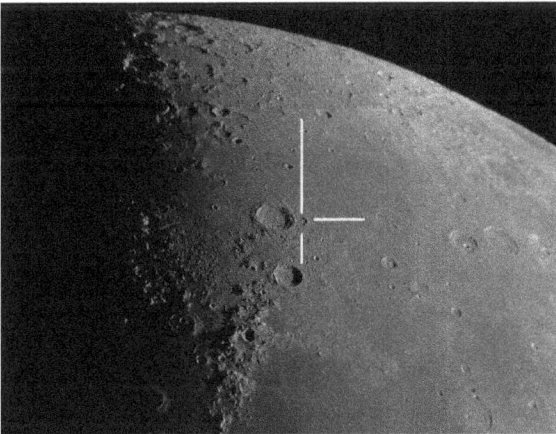

Cassini A

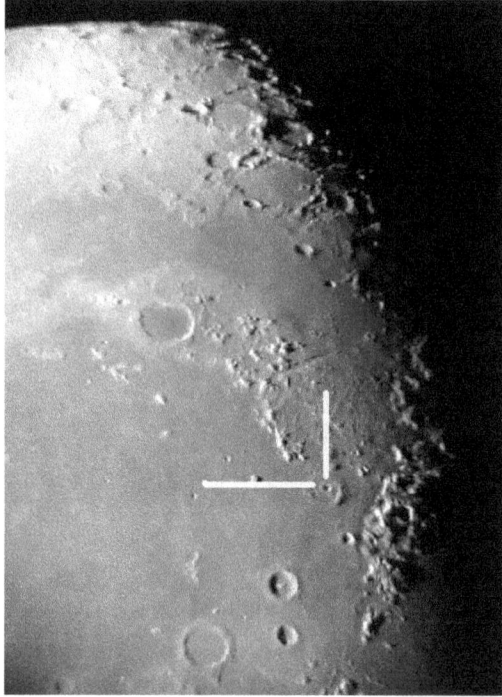

Cassini A is an impact crater within the larger Cassini crater. Cassini is located in the Palus Nebularum, at the eastern end of Mare Imbrium. The floor of Cassini was flooded with lava and the surface has a multitude of impacts, including a pair of significant craters contained entirely within the rim. Cassini A is the larger of these two, and it lies just north-east of the crater center.

TO LOCATE: Find Mare Serenitatis and Mare Imbrium. Look "just over the wall" from Serenitatis and there will be a crater with an off-center interior crater. That will be Cassini with the smaller Cassini A inside.

Manilius

Manilius is an impact crater on the northeast edge of Mare Vaporum. It has a well-defined rim with a sloping inner surface that run directly down to the ring-shaped mound of scree, broken rock that appears at the bottom of crags, mountain cliffs or valley shoulders. The small crater interior has a higher albedo than the surroundings, and it appears bright when the sun is overhead. Within the crater is a central peak formation near the mid-point. The crater also possesses a ray system.

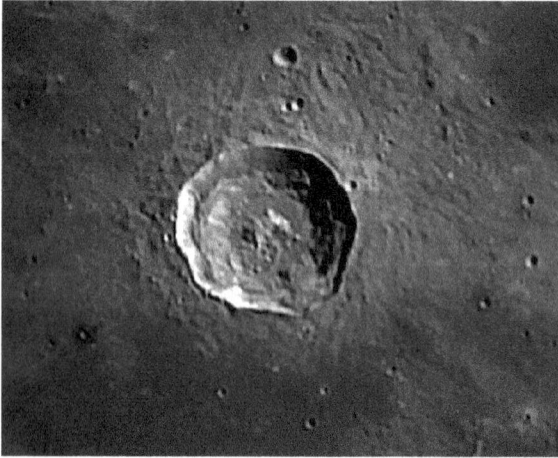

This feature is named for Marcus Manilius, a first century Roman poet, astrologer, and author of a poem in five books called *Astronomica*.

Locate

TO LOCATE: Find Mare Serenitatis and Mare Vaporum. Manilius is found between the two.

Gemma Frisius

Gemma Frisius is a lunar crater that is located in the rugged southern highlands of the Moon. The outer wall of this crater has been heavily damaged by impacts, especially along the north and west sides. The smaller satellite craters D, G, and H are attached to this damaged crater.

The southeastern rim of the crater is also worn, and the inner wall has slumped nearly a third of the distance across the interior floor. The remainder of the floor is relatively level and deep, with a central peak that is offset to the northwest of the mid-point.

The feature is named for Gemma Frisius (1508-1555), a mathematician, cartographer and instrument maker. Frisius became noted for the quality and accuracy of his instruments, which were praised by Tycho Brahe and others. He was the first to describe the method of triangulation still used today in surveying. He was also the first to describe how an accurate clock could be used to determine longitude.

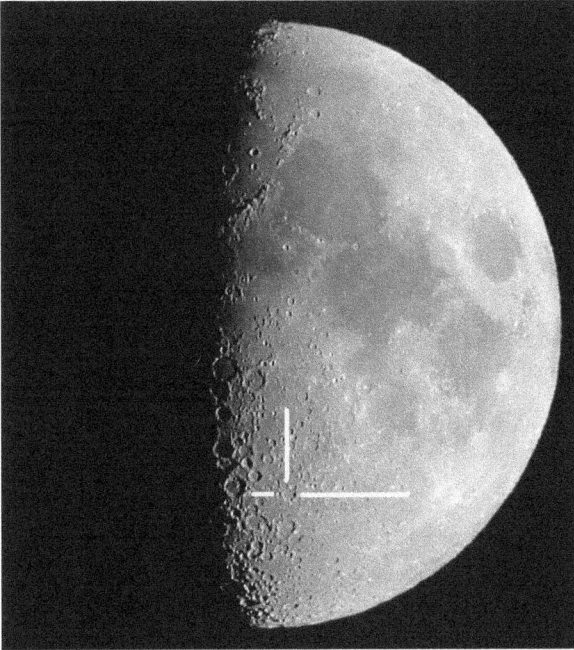

TO LOCATE: Look in the Southern Highlands. We will see what the mind allows us to see. Some observers have noted that this crater has a certain resemblance to a paw print. I, on the other hand, see the face of Mickey Mouse. Compare the image below with that on the preceding page to see the mouse's head.

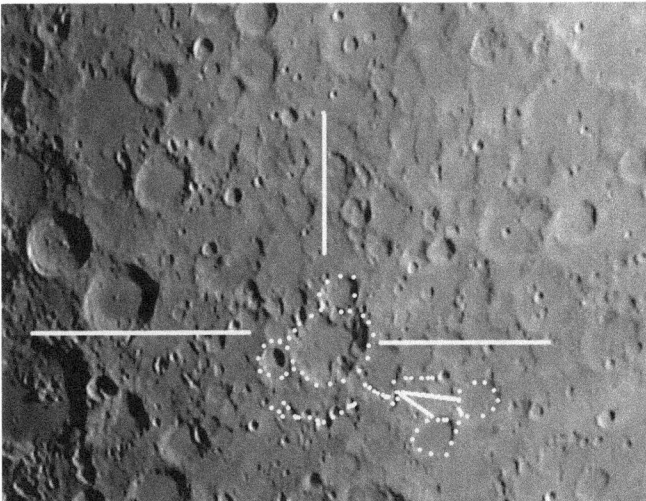

CRATERS OF THE TEN DAY OLD MOON

Visible with a Telescope

Davy

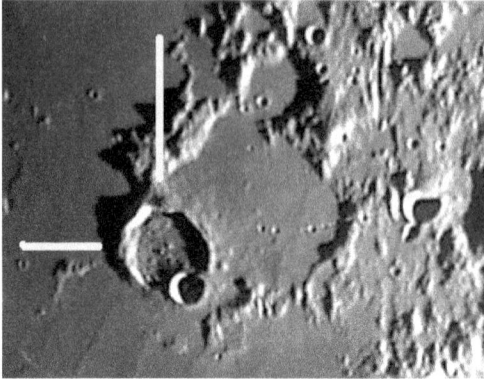

Davy is a small crater located on the eastern edge of the Mare Nubium. It is overlaying the lava-flooded remains of the larger crater to the east, which contains a crater chain designated Catena Davy.

As seen in this photograph taken by Apollo 12 astronauts, Catena Davy forms a fascinating and near perfect line of 23 craters. The chain is likely the cause of a single body that broke apart prior to impact due to tidal effects.

Davy Crater is named for Sir Humphry Davy (1778-1829), a Cornish chemist and physicist. Davy became well known due to his experiments with the physiological action of some gases, including laughing gas (nitrous oxide) - to which he was addicted.

TO LOCATE: Find Mare Nubium. One the edge of Nubium, locate three craters Ptolemaeus, Alphonsus and Arzachel. There is a slight curve to those three craters. Imagine they are an arrow that points (although a bit off center of their target) to Davy.

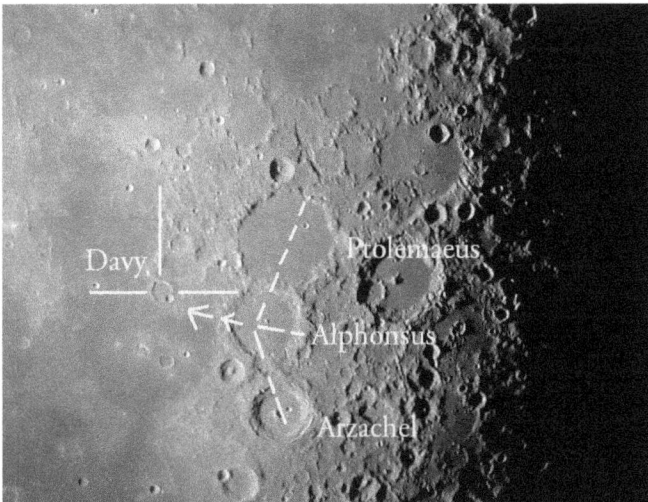

Pitatus

Pitatus is an impact crater at the southern edge of Mare Nubium. Joined to the northwest rim is Hesiodus crater, and the two are joined by a narrow cleft.

Pitatus is a floor-fractured crater, meaning it was flooded from the interior by magma intruding through cracks and openings. The flooded crater floor contains low hills in the east and a system of slender clefts named the *Rimae Pitatus*. The larger and more spectacular of these rilles follow the edges of the inner walls, especially in the northern and eastern halves. The floor also contains the faint traces of deposited ray markings.

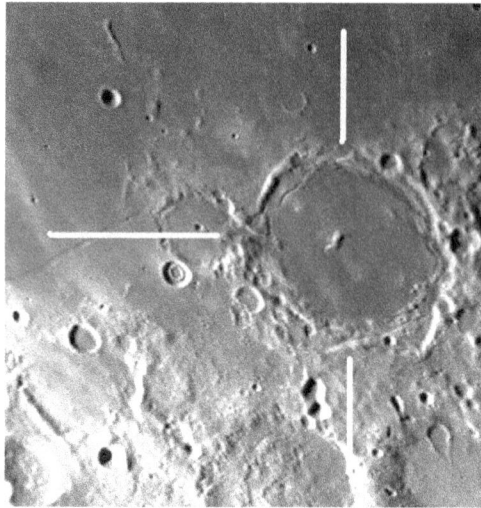

This feature is named for Pietro Pitati (in Latin, Petrus Pitatus), a 16th century Italian astronomer and mathematician.

TO LOCATE: Find Tycho (marked in the box in image below) and move toward Mare Nubium. Pitatus is just inside Nubium.

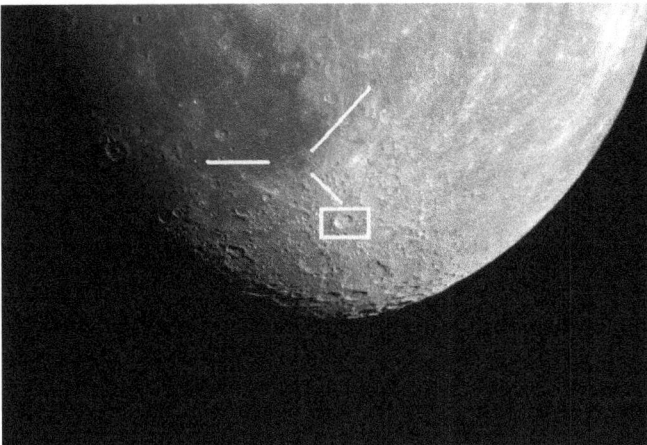

Billy

Billy is a crater located at the southern fringes of the Oceanus Procellarum. It lies to the southeast of the similar-sized Hansteen crater, and west-southwest of the flooded Letronne crater.

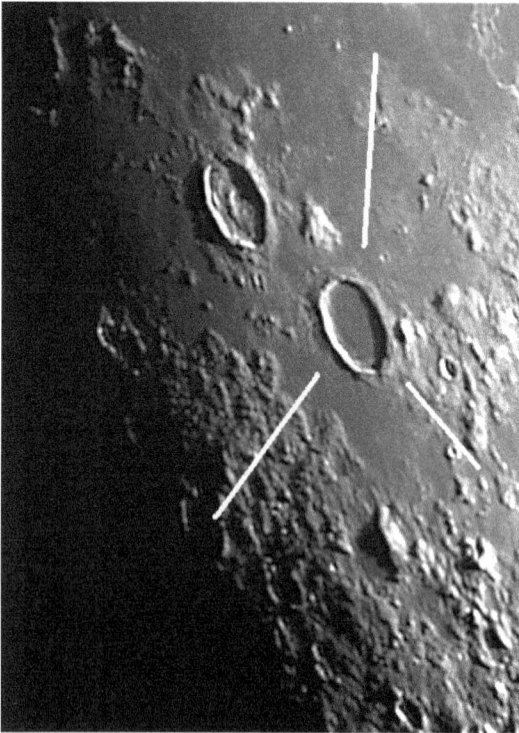

The interior floor of Billy crater has been flooded by basaltic lava, leaving a dark surface due to the low albedo. The portion of the rim remaining above the surface is narrow and low, with a thin inner wall.

This feature is named for Jacques de Billy (1602-1679), a French Jesuit mathematician. Billy was one of the first scientists to reject the role of astrology in science.

TO LOCATE: Find Mare Humorum. Three prominent craters form a triangle – Gassendi, Mersenius and Agatharchides. Draw a line from Agatharchides to Gassendi and beyond to a dark floor crater, and that will be Billy.

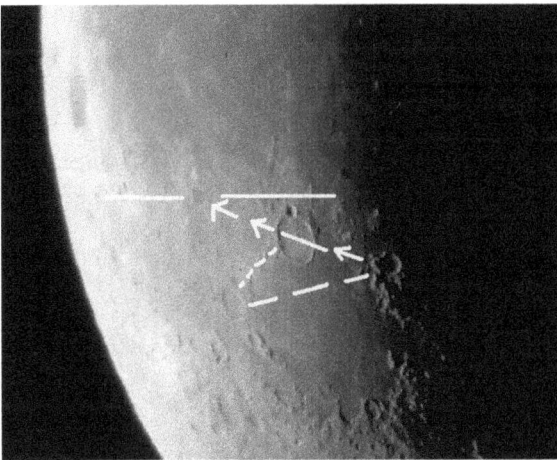

Fra Mauro

Fra Mauro is the worn remnant of a walled lunar plain. It is part of the surrounding Fra Mauro formation located northeast of Mare Cognitum and southeast of Mare Insularum.

The surviving rim of Fra Mauro is heavily worn, with incisions from past impacts and openings in the north and east walls. The rim is the most prominent in the southeast, where it shares a wall with Parry. The remainder consists of little more than low, irregular ridges.

The floor of this formation has been covered by basaltic-lava. This surface is almost divided by clefts running from the north and south rims. There is no central peak.

Just to the north of the walled plain is the landing site of the Apollo 14 mission. The crew sampled breccia[2] that had been deposited here by the Imbrium basin-forming impact, and which partly covers the Fra Mauro crater. This rough debris blanket of ejecta is referred to as the "Fra Mauro Formation".

This feature is named for Fra Mauro, a 15th century Italian monk. He was also a mapmaker, who in 1457 mapped the totality of the Old world with surprising accuracy.

---

[2] Breccia, derived from the Latin word for "broken," is typically a rock composed of angular fragments in a matrix that may be of a similar or a different material.

TO LOCATE: Draw an imaginary arching line from Tycho (marked as 1 in the south), through Bullialdus (marked as 2) and onto Copernicus (marked as 3). Between Bullialdus and Copernicus, in the boxed area highlighted in the photograph above, is Fra Mauro, a very shallow crater bordered by two other slightly smaller craters.

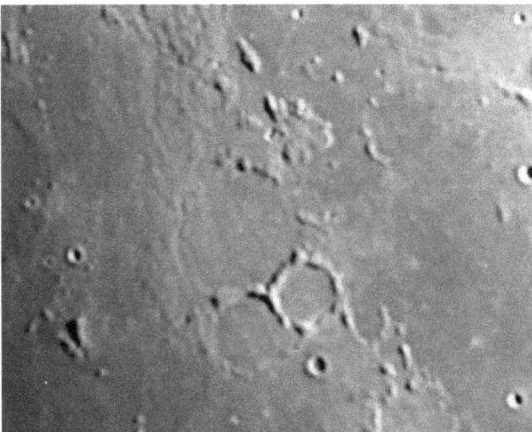

Clavius craterlets

The floor of the crater Clavius forms a plain that is marked by a number of interesting smaller crater impacts. Porter and Rutherford are similar, equal-size craters that intersect the rim of Clavius. The other craters on the floor are Clavius D, C, N, and J, a sequence of diminishing craters that has proved a useful tool for amateur astronomers that want to test the resolution of their small telescopes.

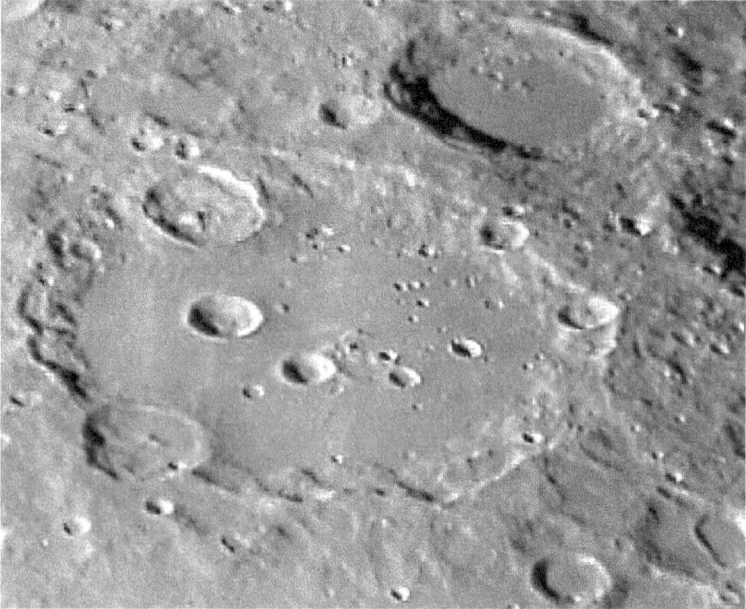

The crater is named for Christopher Clavius, (March 25, 1538 – February 12, 1612), a Jesuit mathematician and astronomer who developed the Gregorian calendar.

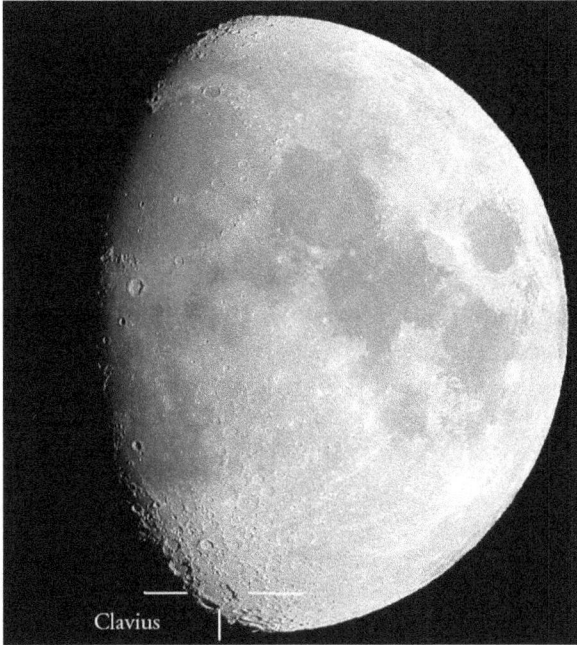

Clavius

TO LOCATE: Clavius appears as a prominent notch in the terminator about 1 to 2 days after the Moon reaches first quarter. It is a prominent and large crater south of the much smaller Tycho.

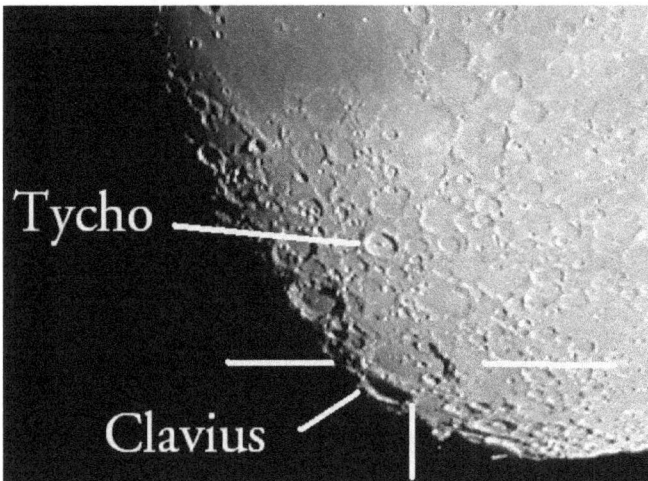

Tycho

Clavius

Hippalus

Hippalus the remnant of a crater on the eastern edge of Mare Humorum. The southwest rim of the crater is missing, and the crater forms a bay along the edge of the mare. The surviving rim is worn and eroded, forming a low, circular mountain range. The lava-flooded floor of Hippalus is bisected by a wide rille belonging to the Rimae Hippalus. This rille follows a course to the south before curving gently to the southwest. The crater floor to the east of this rille is more rugged than the area in the western half.

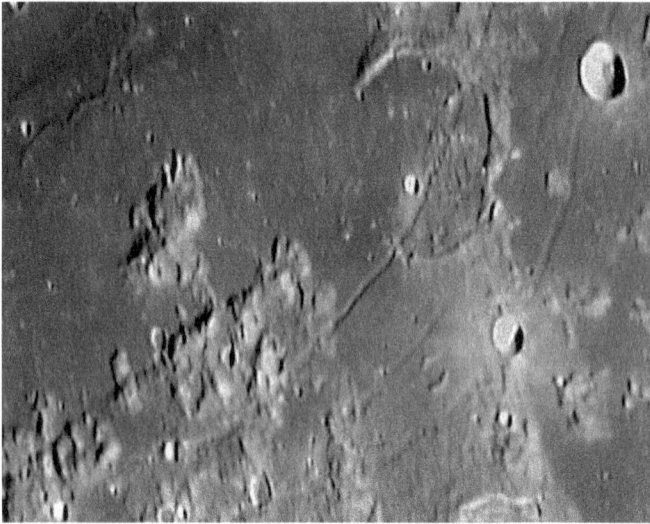

This feature is named for Hippalus, a First Century BC Greek navigator.

TO LOCATE: Hippalus is located on the edge of Mare Humorum. It resembles a letter "C" – which may appear as a reversed letter, depending on your telescope's view.

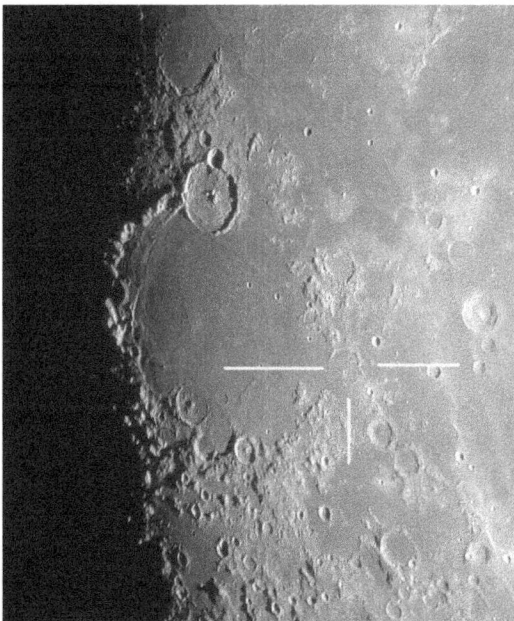

J. Herschel

J. Herschel is large walled-plain crater. It is located in the northern part of the Moon's surface, and appears foreshortened when viewed from the Earth. The southeastern rim of J. Hershel forms part of the edge of Mare Frigoris.

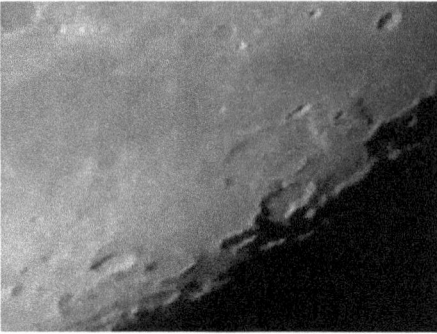

The rim of this crater has been heavily eroded. The remaining rim survives as a ring of ridges that have been reshaped by subsequent impacts. The interior floor is relatively level, but has been marked by a multitude of tiny impacts.

This feature is named for John Frederick William Herschel (1792–1871), an English mathematician and astronomer. He was the son of astronomer William Herschel. John Frederick William Herschel developed the use of the Julian day system in astronomy and made several important contributions to the improvement in the field of photography. In addition to actually coining the term "photography", he discovered sodium thiosulfate as a fixer of silver halides, enabling early photographers to "fix" camera pictures and make them permanent.

TO LOCATE: Notice the point between Sinus Iridium and Mare Imbrium – this is Promontorium Laplace. Imagine that area forming a "V" – continue to repeat the "V" toward the limb until one of these "V"s cradles J. Herschel.

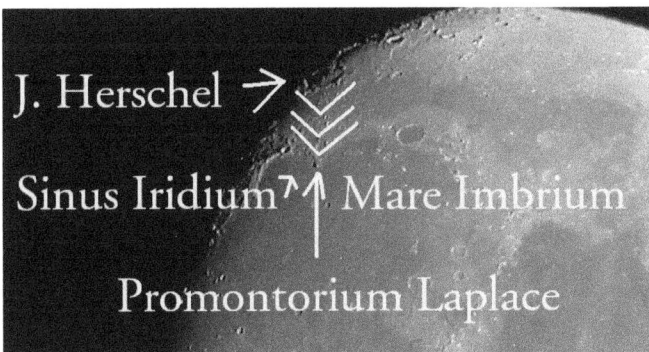

J. Herschel →

Sinus Iridium ↗↑ Mare Imbrium

Promontorium Laplace

CRATERS OF THE FOURTEEN DAY OLD MOON

Visible with a Telescope

Schickard

Schickard is a walled-plain impact crater. It lies in the southwest sector of the moon, near the lunar limb. Because it is so close to the limb, foreshortening causes the crater to appear to be oblong.

The Schickard crater has a worn rim that is overlaid in several locations by smaller impact craters. The most prominent of these is irregular Schickard E crater across the southeast rim.

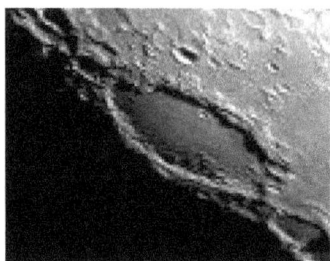

The floor is marked with areas of lighter-albedo material, leaving darker patches in the north and southeast. This feature is more prominent when the sun is at a relatively high angle. There are also multiple small crater impacts on the floor, most notably in the southeast.

This feature is named for Wilhelm Schickard (1592-1635), who built the first automatic calculator in 1623, which contemporaries called a "calculating clock."

TO LOCATE: Tycho, Mare Humorum and Schickard form a triangle with Schickard closes to the lunar limb. Schickard is darker than the surrounding area.

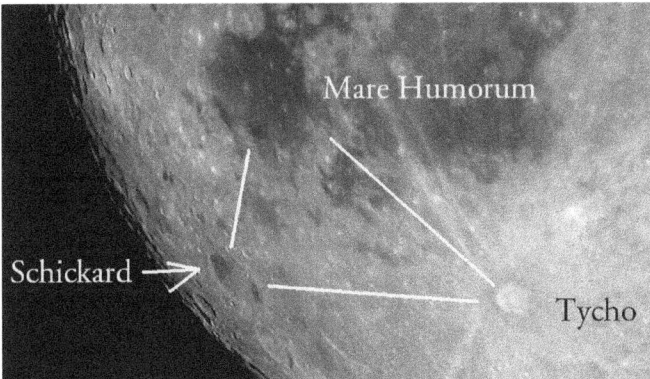

Reiner Gamma

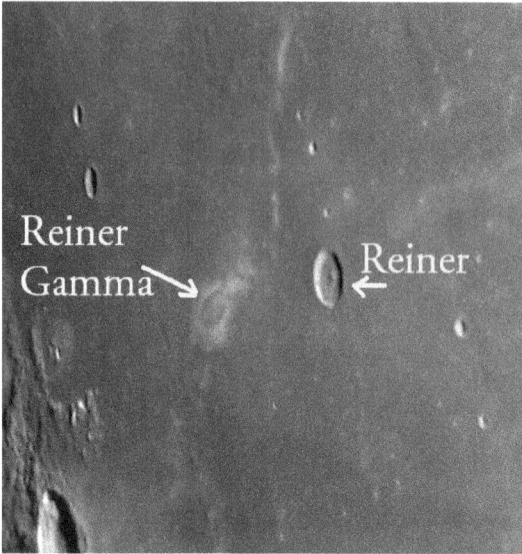

Reiner Gamma is an albedo feature that is located on the Oceanus Procellarum, to the west of the Reiner crater on the Moon. The feature has a higher albedo than the relatively dark mare surface, with a diffuse and oval appearance. The Reiner crater is named for Vincentio Reiner (1606-1647), an Italian astronomer and friend and colleague of Galileo.

Reiner Gamma is one of the strongest localized magnetic anomalies on the Moon. The surface field strength of this feature may be sufficient to form a mini-magnetosphere, which could deflect the solar wind. As the particles in the solar wind are known to darken the lunar surface, the magnetic field at this site may account for the survival of this albedo feature.

TO LOCATE: Imagine a straight line from Copernicus, through Kepler and finally to Reiner (highlighted in the box above.

LUNAR ECLIPSES

An eclipse occurs when body passes into the shadow cast by another body. In astronomy, the best-known type of eclipse occurs whenever the Sun, Earth and Moon line up exactly. If this occurrence is at the time of a full moon where the Moon passes through the Earth's shadow, it is called a lunar eclipse. The type and length of a lunar eclipse depends upon the Moon's location relative to its orbital node. If the lining up of the Sun, Moon and Earth occurs at New Moon, the event is referred to as a solar eclipse.

A **penumbral** eclipse occurs when the Moon only passes through the Earth's penumbra, or the outer portion of the Earth's shadow. The penumbra does not cause the noticeable darkening of the Moon's surface (though some may argue it turns a little yellow). Penumbral eclipses can be

196

partial or total. A total penumbral is a rare type of lunar eclipse.

A **total lunar eclipse** occurs when the Moon travels completely into the Earth's umbra, the dark inner portion of the shadow. The Moon's speed through the shadow is about one kilometer per second, and the totality may last up to nearly 107 minutes. However, the time between the Moon's first contact with the shadow and last contact, when it has completely exited the shadow, may be up to 6 hours and 14 minutes. If only part of the Moon enters the umbra, it is called a **partial lunar eclipse**.

Because the Moon's orbit around the Earth is inclined 5° with respect to the orbit of the Earth around the Sun, lunar eclipses do not occur at every full moon. For an eclipse to occur, the Moon must be near its orbital node —the intersection of the orbital planes. Passing through the shadow at or very close to the node results in a total or partial eclipse.

TABLE OF LUNAR ECLIPSES

Date	Eclipse Type	Geographic Region of Eclipse Visibility
1991 Jan 30	Penumbral	Americas, Europe, w Africa
1991 Jun 27	Penumbral	Americas, s Europe, Africa
1991 Jul 26	Penumbral	e Europe, Africa, Asia, Aus., w Pacific
1991 Dec 21	Partial	Asia, Aus., Pacific, Americas
1992 Jun 15	Partial	e Pacific, Americas, w Europe, Africa
1992 Dec 09	Total	Americas, Europe, Africa, Asia
1993 Jun 04	Total	Asia, Aus., Pacific, w Americas
1993 Nov 29	Total	Pacific, Americas, Europe, w Africa
1994 May 25	Partial	e Pacific, Americas, Europe, Africa
1994 Nov 18	Penumbral	Pacific, Americas, Europe, w Africa
1995 Apr 15	Partial	Asia, Aus., Pacific, w Americas
1995 Oct 08	Penumbral	Europe, Africa, Asia, Aus., Pacific

Date	Eclipse Type	Geographic Region of Eclipse Visibility
1996 Apr 04	Total	Americas, Europe, Africa, w Asia
1996 Sep 27	Total	c Pacific, Americas, Europe, Africa
1997 Mar 24	Partial	c Pacific, Americas, Europe, Africa
1997 Sep 16	Total	Europe, Africa, Asia, Aus.
1998 Mar 13	Penumbral	c Pacific, Americas, Europe, Africa
1998 Aug 08	Penumbral	Americas, Europe, Africa
1998 Sep 06	Penumbral	e Asia, Aus., Pacific, Americas
1999 Jan 31	Penumbral	Europe, Africa, Asia, Aus., Pacific
1999 Jul 28	Partial	e Asia, Aus., Pacific, Americas
2000 Jan 21	Total	Pacific, Americas, Europe, Africa
2000 Jul 16	Total	Asia, Pacific, w Americas

Date	Eclipse Type	Geographic Region of Eclipse Visibility
2001 Jan 09	Total	e Americas, Europe, Africa, Asia
2001 Jul 05	Partial	e Africa, Asia, Aus., Pacific
2001 Dec 30	Penumbral	e Asia, Aus., Pacific, Americas
2002 May 26	Penumbral	e Asia, Aus., Pacific, w Americas
2002 Jun 24	Penumbral	S. America, Europe, Africa, c Asia, Aus.
2002 Nov 20	Penumbral	Americas, Europe, Africa, e Asia
2003 May 16	Total	c Pacific, Americas, Europe, Africa
2003 Nov 09	Total	Americas, Europe, Africa, c Asia
2004 May 04	Total	S. America, Europe, Africa, Asia, Aus.
2004 Oct 28	Total	Americas, Europe, Africa, c Asia
2005 Apr 24	Penumbral	e Asia, Aus., Pacific, Americas
2005 Oct 17	Partial	Asia, Aus., Pacific, North America
2006 Mar 14	Penumbral	Americas, Europe, Africa, Asia

Date	Eclipse Type	Geographic Region of Eclipse Visibility
2006 Sep 07	Partial	Europe, Africa, Asia, Aus.
2007 Mar 03	Total	Americas, Europe, Africa, Asia
2007 Aug 28	Total	e Asia, Aus., Pacific, Americas
2008 Feb 21	Total	c Pacific, Americas, Europe, Africa
2008 Aug 16	Partial	S. America, Europe, Africa, Asia, Aus.
2009 Feb 09	Penumbral	e Europe, Asia, Aus., Pacific, w N.A.
2009 Jul 07	Penumbral	Aus., Pacific, Americas
2009 Aug 06	Penumbral	Americas, Europe, Africa, w Asia
2009 Dec 31	Partial	Europe, Africa, Asia, Aus.
2010 Jun 26	Partial	e Asia, Aus., Pacific, w Americas
2010 Dec 21	Total	e Asia, Aus., Pacific, Americas, Europe
2011 Jun 15	Total	S.America, Europe, Africa, Asia, Aus.
2011 Dec 10	Total	Europe, e Africa, Asia, Aus., Pacific, N.A.
2012 Jun 04	Partial	Asia, Aus., Pacific, Americas

Date	Eclipse Type	Geographic Region of Eclipse Visibility
2012 Nov 28	Penumbral	Europe, e Africa, Asia, Aus., Pacific, N.A.
2013 Apr 25	Partial	Europe, Africa, Asia, Aus.
2013 May 25	Penumbral	Americas, Africa
2013 Oct 18	Penumbral	Americas, Europe, Africa, Asia
2014 Apr 15	Total	Aus., Pacific, Americas
2014 Oct 08	Total	Asia, Aus., Pacific, Americas
2015 Apr 04	Total	Asia, Aus., Pacific, Americas
2015 Sep 28	Total	e Pacific, Americas, Europe, Africa, w Asia
2016 Mar 23	Penumbral	Asia, Aus., Pacific, w Americas
2016 Aug 18	Penumbral	Aus., Pacific, Americas
2016 Sep 16	Penumbral	Europe, Africa, Asia, Aus., w Pacific
2017 Feb 11	Penumbral	Americas, Europe, Africa, Asia
2017 Aug 07	Partial	Europe, Africa, Asia, Aus.
2018 Jan 31	Total	Asia, Aus., Pacific, w N.America

Date	Eclipse Type	Geographic Region of Eclipse Visibility
2018 Jul 27	Total	S.America, Europe, Africa, Asia, Aus.
2019 Jan 21	Total	c Pacific, Americas, Europe, Africa
2019 Jul 16	Partial	S.America, Europe, Africa, Asia, Aus.
2020 Jan 10	Penumbral	Europe, Africa, Asia, Aus.
2020 Jun 05	Penumbral	Europe, Africa, Asia, Aus.
2020 Jul 05	Penumbral	Americas, sw Europe, Africa
2020 Nov 30	Penumbral	Asia, Aus., Pacific, Americas
2021 May 26	Total	e Asia, Australia, Pacific, Americas
2021 Nov 19	Partial	Americas, n Europe, e Asia, Australia, Pacific
2022 May 16	Total	Americas, Europe, Africa
2022 Nov 08	Total	Asia, Australia, Pacific, Americas
2023 May 05	Penumbral	Africa, Asia, Australia
2023 Oct 28	Partial	e Americas, Europe, Africa, Asia, Australia
2024 Mar 25	Penumbral	Americas

Date	Eclipse Type	Geographic Region of Eclipse Visibility
2024 Sep 18	Partial	Americas, Europe, Africa
2025 Mar 14	Total	Pacific, Americas, w Europe, w Africa
2025 Sep 07	Total	Europe, Africa, Asia, Australia
2026 Mar 03	Total	e Asia, Australia, Pacific, Americas
2026 Aug 28	Partial	e Pacific, Americas, Europe, Africa
2027 Feb 20	Penumbral	Americas, Europe, Africa, Asia
2027 Jul 18	Penumbral	e Africa, Asia, Australia, Pacific
2027 Aug 17	Penumbral	Pacific, Americas
2028 Jan 12	Partial	Americas, Europe, Africa
2028 Jul 06	Partial	Europe, Africa, Asia, Australia
2028 Dec 31	Total	Europe, Africa, Asia, Australia, Pacific
2029 Jun 26	Total	Americas, Europe, Africa, Mid East
2029 Dec 20	Total	Americas, Europe, Africa, Asia
2030 Jun 15	Partial	Europe, Africa, Asia, Australia

Date	Eclipse Type	Geographic Region of Eclipse Visibility
2030 Dec 09	Penumbral	Americas, Europe, Africa, Asia
2031 May 07	Penumbral	Americas, Europe, Africa
2031 Jun 05	Penumbral	East Indies, Australia, Pacific
2031 Oct 30	Penumbral	Americas
2032 Apr 25	Total	e Africa, Asia, Australia, Pacific
2032 Oct 18	Total	Africa, Europe, Asia, Australia
2033 Apr 14	Total	Europe, Africa, Asia, Australia
2033 Oct 08	Total	Asia, Australia, Pacific, Americas
2034 Apr 03	Penumbral	Europe, Africa, Asia, Australia
2034 Sep 28	Partial	Americas, Europe, Africa
2035 Feb 22	Penumbral	e Asia, Pacific, Americas
2035 Aug 19	Partial	Americas, Europe, Africa, Mid East
2036 Feb 11	Total	Americas, Europe, Africa,, Asia, w Australia
2036 Aug 07	Total	Americas, Europe, Africa, w Asia

Date	Eclipse Type	Geographic Region of Eclipse Visibility
2037 Jan 31	Total	e Europe, e Africa, Asia, Australia, Pacific, N.A.
2037 Jul 27	Partial	Americas, Europe, Africa
2038 Jan 21	Penumbral	Americas, Europe, Africa
2038 Jun 17	Penumbral	e N. America, C. & S. America, Africa, w Europe
2038 Jul 16	Penumbral	Australia, e Asia, Pacific, w Americas
2038 Dec 11	Penumbral	Europe, Africa, Asia, Australia
2039 Jun 06	Partial	Europe, Africa, Asia, Australia
2039 Nov 30	Partial	Europe, Africa, Asia, Australia, Pacific
2040 May 26	Total	e Asia, Australia, Pacific, w Americas
2040 Nov 18	Total	e Americas, Europe, Africa, Asia, Australia

Geographic abreviations (used above): n = north, s = south, e = east, w = west, c = central

A Lunar Glossary

A

Absolute altitude, The height of any point on the Lunar surface in comparison to the "reference sphere," a perfect sphere of 3476 kilometers in diameter which represents the mean height of average terrain on the Moon. The necessity of this is that there is no sea-level, as there is on earth, to measure altitude of mountains.

Albedo, A measure of reflectiveness or reflective power, or, specifically, the light that is reflected by the surface of a body such as the Moon. The term **"low albedo"** generally refers to dark features; **"high albedo"** generally refers to lighter-colored features.

Anorthosite, Granular igneous rock usually comprised of soda-lime feldspar.

Apogee, The point in the Moon's orbit where it is furthest from Earth. At its apogee, the Moon is 406,700 kilometers from Earth. (See also *__perigee__*.)

Apollo, Name given to the United States manned Lunar program, which included the first successful landing of a human crew on the Moon (*Apollo 11*, 20 July 1969).

B

Basin, A large impact _crater_, usually with a diameter in excess of 100 kilometers. Most basins have been modified by degradation of the original basin relief through downslope movement of debris and flooding of the basin interior by lavas.

Breccia, Coarse-grained rock composed of angular fragments of pre-existing rock.

C

Caldera, A large volcanic depression at the summit of a volcano, caused by collapse or explosion.

Catena _pl._ **catenae**, Chain of craters.

Cavus _pl._ **cavi**, Hollows or irregular steep-sided depressions, usually in arrays or clusters.

Chasma _pl._ **chasmata**, A deep, elongated, steep-sided depression.

Colles, Small hills or knobs.

Crater _pl._ **craters**, A typically bowl-shaped or saucer-shaped pit or circular depression, generally of considerable size and with steep inner slopes, formed on a surface or in the ground by the explosive release of chemical or kinetic energy; e.g., an "**impact crater**" or an "**explosion crater**".

D

Diurnal, Having a daily cycle, or recurring every day.

Dorsum *pl.* **dorsa**, Ridge.

E

Ejecta, The material thrown out of an impact crater by the shock pressures generated during the impact event. Ejecta generally covers the surface around an impact crater to a distance of at least one crater diameter, with individual streamers of material extending well beyond this distance. The ejecta blanket of a crater becomes less visible with increasing age of the crater. (See also *rays*)

F

Facula *pl.* **faculae**, Bright spot.

Far side, The surface of the Moon that is not generally visible from Earth due to its unique orbital pattern, which keeps the Lunar "face" turned toward Earth.

Farrum *pl.* **farra**, Pancake-like structure, or a row of such structures.

Flexus, A very low curvilinear ridge with a scalloped pattern.

Fluctus, Flow terrain.

Fossa *pl.* **fossae**, Long, narrow, shallow depression.

Full moon, Lunar phase during which the entire visible surface is under illumination. (See also ***new moon***.)

G

Gibbous moon, The phase of the Moon during which more than half, but less than all, the visible hemisphere of the Moon is illuminated by sunlight.

H

Highlands, The densely cratered portions of the Moon that are typically at higher elevations than the mare plains; often referred to as "***terrae***." The highlands contain a significant proportion of ***anorthosite***, an igneous rock made up almost entirely of plagioclase feldspar.

L

Labes, Landslide.

Labyrinthus *pl.* **labyrinthi**, Complex of intersecting valleys.

Lacus, "Lake"; small plain.

Large ringed feature, Unusual ringed features on the lunar surface that cannot be classified under another descriptor.

Lava, Volcanic rock extruded by the eruption of molten material. An extensive segment of the lunar surface, specifically in the **_mare_** regions, is comprised primarily of basalt resulting from lava flows.

Limb, The outer edge of a Lunar or other planetary disk.

Linea *pl.* **lineae**, A dark or bright elongate marking; may be curved or straight.

Luna, Accepted common name for Earth's Moon; derived from the Latin word for "light."

Lunar eclipse, Period in which the Earth is positioned so as to obscure the Moon from sunlight.

Lunation, The period of time it takes the Moon to complete one set of phases (the "**_synodic month_**"), specifically from New Moon to New Moon, averaging 29 days, 12 hours, 44 minutes and 2.9 seconds. Lunations are numbered sequentially, beginning with Lunation 1 which commenced on 16 January 1923. Lunation 1000 will commence on 25 October 2003.

M

Macula *pl.* **maculae**, Dark spot; may be irregular.

Mare (pronounced "mahr-ay") *pl.* **maria** (pronounced "ma-ree-ah"), "Sea"; a large circular plain on the Moon; specifically, the low **_albedo_** plains covering the floors of several large **_basins_** and spreading over adjacent areas. The mare material is comprised primarily of basaltic ***lava*** flows, in contrast to the ***anorthosites*** in the highlands.

Mascon, Concentrations of mass on the lunar surface (from *mass con*centrations).

Massif, A massive topographic and structural feature, commonly formed of rocks more rigid than those of its surroundings. These rocks may be protruding bodies of basement rocks, consolidated during earlier ***orogenies***.

Mensa *pl.* **mensae**, A flat-topped prominence with cliff-like edges.

Mons *pl.* **montes**, Mountain.

N

New moon, lunar phase during which the entire visible surface is in darkness. (See also "***full moon***.")

O

Oceanus, A very large dark area on the Moon.

P

Palus *pl.* **paludes**, "Swamp"; small plain.

Patera *pl.* **paterae**, An irregular crater, or a complex one with scalloped edges.

Perigee, The point in the Moon's orbit where it is closest to Earth. At its perigee, the Moon is 356,400 kilometers from Earth. (See also *apogee*.)

Planitia *pl.* **planitiae**, Low plain.

Planum *pl.* **plana**, Plateau or high plain.

Promontorium *pl.* **promontoria**, A high point of land; headland.

R

Ray, A streamer of *ejecta* associated with an impact crater. Rays are most often of higher *albedo* than their surroundings. The albedo contrast may result from either disruption of the local surface by the ejecta or by emplacement of ejecta on the surroundings, or both.

Regio *pl.* **regiones**, A large area marked by reflectivity or color distinctions from adjacent areas; a broad geographic region.

Regolith, A residual mixture of fine dust and rocky debris, usually produced by meteor impacts, covering the lunar surface.

Rille, One of the several trench-like or crack-like valleys, up to several hundred kilometers long and one to two kilometers wide, commonly occurring on the Moon's surface. Rilles may be extremely irregular with meandering courses ("**sinuous rilles**"), or they may be relatively straight ("**normal**

rilles"); they have relatively steep walls and usually flat bottoms. Rilles are essentially youthful features and apparently represent fracture systems originating in brittle material.

Rima *pl.* **rimae**, Fissure.

Rupes, Scarp.

S

Scarp, A change in topography along a linear to arcuate cliff. The cliff may be the result of one or more processes including tectonic, volcanic, impact-related, or degradational processes. The term "*__rupes__*" is generally used in lunar geography when referring to this type of feature.

Scopulus, Lobate or irregular scarp.

Secondary craters, Craters produced by the impact of debris thrown out by a large impact event. Many secondary craters occur in clusters or lines where groups of ejecta blocks impacted almost simultaneously.

Selene (pronounced "suh-lee-nee"), The Greek goddess of the Moon.

Selenology, The scientific study of the history of the Moon, as recorded in rocks, minerals and other materials found on the lunar surface; from Selene, Greek goddess of the Moon.

Sinus, "Bay"; small plain.

Sulcus *pl.* **sulci**, Subparallel furrows and ridges.

Synodic month or **synodic period**, The period of time it takes for one body to orbit around its primary, such as the Moon around Earth. The Lunar synodic month is measured as the time it takes to complete one set of phases, specifically from New Moon to New Moon. (See also ***lunation***.)

T

Terminator, The line separating the illuminated and dark areas of a planetary body; the dividing line between day and night as observed from a distance.

Terra *pl.* **terrae**, Extensive land mass; often used as descriptor for lunar highlands.

Tessera *pl.* **tesserae**, Tile-like, polygonal terrain.

Tholus *pl.* **tholi**, Small domical mountain or hill.

U

Undae, Dunes.

V

Vallis *pl.* **valles**, Valley.

Vastitas *pl.* **vastitates**, Extensive plain.

W X Y Z

Waning moon, Period during which illumination of the visible lunar surface decreases (following the most recent _**full moon**_) until it reaches complete darkness ("_**new moon**_").

Waxing moon, Period during which illumination of the visible lunar surface increases (following the most recent _**new moon**_) until it reaches complete illumination ("_**full moon**_").

Lunar Club Program
Of The
Astronomy League

This atlas follows the various features on the moon that are identified in the Astronomy League's Lunar Observing Program. Whether or not you are a member of the League, following the challenges of this program can enable you to gain an understanding of the lunar geography. On the pages that follow are the charts that correspond with the program. As you find each feature, note the date and time, along with other comments you may wish to make, such as what instrument was used to observe the feature.

Naked Eye Objects

Object	Date & Time	Comments	See Page
(Within 72 Hrs of new) Old Moon in New Moon's Arms			9
(Within 72 Hrs of new) New Moon in Old Moon's Arms			9
(Within 40 Hrs of new) Crescent Moon, Waxing			
(Within 48 Hrs of New) Crescent Moon, Waning			
Man in the Moon			10
Woman in the Moon			10
Rabbit in the Moon			11
Cow Jumping Over the Moon			11

Naked Eye Objects - Maria

Object	Date & Time	Comments	See Page
Crisium			14
Fecunditatis			15
Serenitatis			16
Tranquillitatis			17
Nectaris			18
Imbrium			19
Frigoris			20
Nubium			21
Humorum			22
Oceanus Procellarum			23

Binocular Objects - Features

Object	Date & Time	Comments	See Page
Lunar Rays			26
Sinus Iridum			28
Sinus Medii			29
Sinus Roris			30
Palus Somnii			31
Palus Epidemiarum			32
Mare Vaporum			33

Binocular Objects – Craters of the 4-Day Moon

Object	Date & Time	Comments	See Page
Langrenus			36
Vendelinus			38
Petavius			40
Cleomedes			42
Atlas			44
Hercules			46
Endymion			48
Macrobius			50

Binocular Objects – Craters of the 7-Day Moon

Object	Date & Time	Comments	See Page
Piccolomini			54
Theophilus			56
Cyrillus			58
Catharina			60
Posidonius			62
Fracastorius			64
Aristoteles			66
Eudoxus			68
Cassini			70
Hipparchus			72
Albategnius			74
Aristillus			76
Autolycus			78
Maurolycus			80

Binocular Objects – Craters of the 10-Day Moon

Object	Date& Time	Comments	See Page
Plato			84
Archimedes			86
Ptolemaeus			88
Alphonsus			90
Arzachel			92
Walter			94
Maginus			98
Tycho			96
Clavius			100
Eratosthenes			106
Longomontanus			104
Copernicus			102
Bullialdus			108
Aristarchus			110
Gassendi			112

Binocular Objects – Craters of the 14-Day Moon

Object	Date & Time	Comments	See Page
Kepler			**116**
Grimaldi			**118**

Additional Comments on Binocular Objects:

Telescopic Objects – Features

Object	Date & Time	Comments	See Page
Sinus Aestuum			122
Lacus Mortis			123
Palus Putredinis			124
Promontorium Laplace			125
Promontorium Heraclides			126
Promontorium Agarum			127
Montes Alpes			128
Montes Apenninus			129
Mons Hadley			130
Mons Piton			132
Mons Pico			134
Rupes Altai			136
Rima Hyginus			138
Vallis Schroteri			140
Vallis Alpes			142
Rupes Recta (straight wall)			144

Telescopic Objects – Craters of the 4-Day Moon

Object	Date & Time	Comments	See Page
Picard			**148**
Furnerius			**150**
Petavius Wall			**152**
Messier Messier A			**154**
Proclus			**156**
Fabricius			**158**

Telescopic Objects – Craters of the 7-Day Moon

Object	Date & Time	Comments	See Page
Plinius			**162**
Mitchell			**164**
Cassini A			**166**
Manilius			**168**
Gemma Frisius			**170**

Telescopic Objects – Craters of the 10-Day Moon

Object	Date & Time	Comments	See Page
Davy			174
Pitatus			176
Billy			178
Fra Mauro			180
Clavius craterlets			182
Hippalus			184
Herschel, J.			186

Telescopic Objects – Craters of the 14-Day Moon

Object	Date & Time	Comments	See Page
Schickard			**190**
Reiner Gamma			**192**

Additional Comments on Telescopic Objects:

Index

Catharina	**60**
Clavius	**100**
Clavius craterlets	**182**
Cleomedes	**42**
Copernicus	**102**
Crisium	**14**
Cyrillus	**58**
Davy	**174**
Endymion	**48**
Eratosthenes	**106**
Eudoxus	**68**
Fabricius	**158**
Fracastorius	**64**
Fra Mauro	**180**

Fecunditatis	**15**
Frigoris	**20**
Furnerius	**150**
Gassendi	**112**
Gemma Frisius	**170**
Grimaldi	**118**
Hercules	**46**
Herschel, J.	**186**
Hippalus	**184**
Hipparchus	**72**
Humorum	**22**
Imbrium	**19**
Kepler	**116**

Lacus Mortis	123
Langrenus	36
Longomontanus	104
Lunar Rays	26
Macrobius	50
Maginus	98
Manilius	168
Maurolycus	80
Messier Messier A	154
Mitchell	164
Mons Hadley	130
Mons Pico	134
Mons Piton	132

Montes Alpes	**128**
Montes Apenninus	**129**
Nectaris	**18**
Nubium	**21**
Oceanus Procellarum	**23**
Palus Epidemiarum	**32**
Palus Putredinis	**124**
Palus Somnii	**31**
Petavius	**40**
Petavius Wall	**152**
Picard	**148**
Piccolomini	**54**
Pitatus	**176**
Plinius	**162**

Plato	84
Posidonius	62
Proclus	156
Promontorium Agarum	127
Promontorium Heraclides	126
Promontorium Laplace	125
Ptolemaeus	88
Reiner Gamma	192
Rima Hyginus	138
Rupes Altai	136
Rupes Recta (straight wall)	144
Schickard	190
Serenitatis	16
Sinus Aestuum	122
Sinus Iridum	28

Sinus Medii	29
Sinus Roris	30
Theophilus	56
Tranquillitatis	17
Tycho	96
Vallis Alpes	142
Vallis Schroteri	140
Vaporum	33
Vendelinus	38
Walter	94

To order additional copies of this book,
contact the author, Maynard Pittendreigh
at
www.Pittendreigh.com

www.ingramcontent.com/pod-product-compliance
Lightning Source LLC
Chambersburg PA
CBHW031926190326
41519CB00007B/423

9 780615 135281